MALADAPTATION

MALADAPTATION

NATURAL SELECTION IN THE WRONG DIRECTION?

PHILIP G. MADGWICK

OXFORD
UNIVERSITY PRESS

OXFORD
UNIVERSITY PRESS

Great Clarendon Street, Oxford, OX2 6DP,
United Kingdom

Oxford University Press is a department of the University of Oxford.
It furthers the University's objective of excellence in research, scholarship,
and education by publishing worldwide. Oxford is a registered trade mark of
Oxford University Press in the UK and in certain other countries

Published in the United States of America by Oxford University Press
198 Madison Avenue, New York, NY 10016, United States of America

British Library Cataloguing in Publication Data
Data available

Library of Congress Control Number: 2024907426

ISBN 9780192870469

DOI: 10.1093/9780191966767.001.0001

Printed and bound by
CPI Group (UK) Ltd, Croydon, CR0 4YY

Preface

Natural selection is expected to improve the ability of organisms to survive and reproduce. Against this expectation, the scientific argument of this book is that natural selection can also act in the opposite direction. The resulting traits are synthesised together as 'maladaptations', giving this neglected and confusing term a precise meaning as a trait that decreases an individual's fitness. The overarching aim of the book is to provide the first steps towards establishing the possibility, reality, and importance of maladaptation in the design of living things.

This book is written for a wide audience, but with three types of reader especially in mind. The first are those with a general interest in science and the natural world. I have endeavoured to write the book in an accessible style to communicate the scientific argument as it is, such that any inquisitive reader should feel empowered to come to their own conclusions. I have not tried to shy away from the scientific details of my argument. Above all, my goal for these readers is to convey the big picture. The book follows a classic hourglass-shaped argument, from the big picture to the nitty-gritty details, and back to the big picture again, and it may be that a general reader finds the middle chapters on the problems of finding evidence for maladaptation somewhat less interesting. I would encourage a general reader to whizz through any dull sections to focus on grasping the change in outlook that comes with identifying traits as maladaptations. In the later chapters, I do more entertainingly venture beyond the confines of current evidence into more speculative questions about how much the maladaptive failings of living things have been normalised in our imaginations as unavoidable 'facts of life'. The argument urges us to look at the natural world with fresh eyes, setting our imaginations free to better understand ourselves and the world around us—and even help to change it.

The second readers are students studying a biological science. In the current mainstream of evolutionary biology, there are a bewildering variety of 'perspectives' on how evolution by natural selection works: Darwinian individualism, the gene's-eye view of evolution, multi-level selection, and so on. My goal is not to help a student figure out which perspective they prefer, because any of them can be useful to address particular puzzles when they are correctly formulated. Instead, my goal is to persuade that, to be correctly formulated, each perspective must be reconciled with the population biology of evolution by natural selection. Indeed, it is only through the integrated understanding of the evolution, genetics, *and ecology* of populations that maladaptation can become comprehensible. Building largely on the perspective from the gene's-eye view of evolution that crystallises the insights of integrating evolution and genetics, I argue that the actions of genes need to be situated in their wider context. This book is a much-needed triumph for a reformed population biology, identifying many unanswered questions for curious minds to address going forward.

The third readers are experts in evolutionary biology. I believe that my definition of maladaptation provides a concept for a kind of design imperfection that lots of evolutionary biologists have been talking around for many years. But, it is clear from the proposal and chapter reviews that the thesis will provoke a diversity of expert responses, including those that, despite fully understanding my argument, are against the possibility, reality, and/or importance of maladaptation to organism design. To those readers, I commend that the book is not intended to be an encyclopaedic review of examples of maladaptation, which I think would be of as much use as a similar review of adaptation. My goal for expert readers is to establish some questions of interest and the problems that I can foresee for those that would seek to answer them. Such restraint has been a source of frustration for some of my supportive reviewers, as I have been reticent to strongly advocate any one example as a clear-cut maladaptation. It remains my belief that definitive evidence of maladaptation would require data that existing studies have never intentionally sought, and so every current example that I know demands greater study. My hope is that the theory of maladaptation will inspire further development and the evidential challenges will be addressed with experimental ingenuity that I can yet scarcely fathom.

Lastly, I would like to thank the many people who have supported me on the road to publication. This book has been enabled by personal and societal adversity, in the work behind unobtained fellowships and the boredom of COVID-19 lockdowns. Much has changed in my life since I started to work on this book. I must whole-heartedly thank the enduring and unsparing support of my family, friends, and colleagues. In particular, I would like to acknowledge Arvid Ågren for his practical encouragements in book-writing; Ian Sherman, Janine Fisher, and Katie Lakina at Oxford University Press for their enthusiastic support; all eight anonymous referees of the book proposal; and the many wonderful people who have commented on the manuscript, especially Elizabeth Madgwick, Christopher Milne, Thomas White, Emily Stevens, and Mariabeth Silkey.

Philip G. Madgwick, Woking, UK

Contents

I

Making room for maladaptation

From the 'rudimentary' wings of a flying squirrel to the 'extreme perfection' of the human eye, the natural world is saturated with the appearance of design. This presents us with a fascinating puzzle: why do living things appear designed, and what are they designed for? Whilst these questions necessarily tread on the toes of philosophers and theologians, there is a perfectly scientific account that neither can reasonably deny. Like any other feature of nature, the appearance of design is explicable in terms of natural processes that are acting all around us. Discerning how these processes work is not an easy task because life is mind-bogglingly complex; there is an extraordinary variety of form and behaviour within nature, and modern biology has only begun to understand the mechanistic details of how this spectacular show comes about. Yet, through the painstaking accumulation of evidence, a great deal is now understood about how living things have come to bear the hallmarks of design.

The appearance of design across nature is explained through evolution by natural selection, which can be understood in strikingly simple terms. Wherever individual organisms display heritable variation in their ability to survive and reproduce, the next generation tends to be made up from individuals that are slightly better at surviving and reproducing. Such statements have been thought to be so trivial as to be tautologous, but the inheritance of traits from individuals in one generation to the next ensures a gradual evolution. The power of the evolutionary process has been most persuasively verified through selective breeding, where a human chooses which individuals are the parents of the next generation. Indeed, this was the evidence that Charles Darwin (1859) relied upon when he first presented the case for natural selection in *The Origin of Species*, supported by a near-encyclopaedic catalogue of variation among domesticated stocks. A staggering diversity of plant and animal breeds have been produced by humans over hundreds of years, leading to diversity in morphology as extreme as that between a Brussels sprout and a cauliflower, or diversity in temperament as extreme as between a Yorkshire Terrier and a Newfoundland. In nature the results are even more impressive, but there are no breeders to choose which individuals successfully reproduce. Instead, as Darwin explained, the environment takes on the role of the human breeder in choosing which individuals preferentially survive and reproduce through a natural selection over millions of years.

Although this simple presentation of the principle of natural selection is coherent, it is not a watertight argument. Reconsidering the results from human feats of selective breeding, alongside the astonishing diversity of traits that have been brought together within breeds on a human whim, there are also many other selected traits that no breeder would ever desire. For instance, over 70% of Cavalier King Charles Spaniels develop a skull malformation that reduces the space available for their brain, leading to syringomyelia, which can develop into chronic pain and even paralysis (Parker et al. 2011). There are also traits that are detrimental to individuals that are not regularly referred to as diseases because they are invariant for all individuals. Take dwarf rabbit breeds such as the Netherland Dwarf, where 25% of each litter contains malformed individuals that die within a few days of birth (Carneiro et al. 2017). There are a large number of other harmful traits in domesticated breeds, but also in natural populations, including the human species. Whilst particular hereditary diseases like cystic fibrosis, haemophilia, or phenylketonuria are rare across the whole of humanity, as a collective, as many as 446 million people are thought to suffer from identifiable hereditary diseases across the world (Wakap et al. 2020). In recent years, more phenomena have come to be recognised as diseases, such as autoimmune disorders, which are especially prevalent in women (Wang et al. 2015). The existence of these harmful traits might cause some confusion about what is going on with evolution, but Darwin (1859, p. 211) was insistent that 'Natural selection will never produce in a being any structure more injurious than beneficial to that being, for natural selection acts solely by and for the good of each'. So, from syringomyelia to autoimmunity, has natural selection gone wrong? How can harmful traits persist even though they are detrimental to individuals? And what does all this tell us about design in the natural world?

In contrast to previous explanations of design in nature through the divine creation of fixed species, there is a static inefficiency that arises from evolutionary change that might explain harmful traits. Under evolution by natural selection, new traits must be modified incrementally out of existing traits through 'tinkering' (Jacob 1977, p. 1161), where some designs are favoured over others on the basis of slight modifications. Indeed, Darwin (1859, p. 181–182), in discussing transitional forms of wing-like structures, compares the 'less perfectly gliding squirrels' to bats that still bear 'traces of an apparatus originally constructed for gliding' and birds that 'have acquired their perfect power of flight'. Despite a largely innocuous role in the evolution of wings, tinkering can introduce inefficiency in other cases, which is famously clear for the recurrent laryngeal nerve (Ridley 1993, p. 60–61). In fish, the nerve travels from the voice-box to the brainstem by the most direct route, passing under the aortic arch of the heart. In modern-day mammals, which are descended from fish, the nerve takes the same route, but this route is circuitous because of the subsequent evolution of the neck. In most mammals, the nerve is only wastefully longer by a few extra centimetres. In giraffes, the evolution of an extremely long neck has meant that the recurrent laryngeal nerve performs a ludicrously wasteful meander to extend for nearly five metres longer than it needs to. Such anatomical acrobatics arises as an inefficiency from tinkering because there is no rational oversight of organism design to reorganise the connection. In the same way, it is conceivable that harmful traits could represent an inefficiency borne out of the incremental modification of evolving designs. This may explain the risk of some diseases like appendicitis, which arises in humans because of a vestigial organ that is no longer functional, but it cannot explain many examples like those already discussed.

If not a straight-forward inefficiency of evolving design, harmful traits could be a symptom of a dynamic inefficiency in evolution by natural selection. Within a generation, natural selection plays a destructive role in eliminating trait variation in the population to favour the best design. For natural selection to play a creative role through accumulating incremental improvements in design over the generations, there must be a continual source of new variation to fuel natural selection into the future. The theory of evolution by natural selection does not depend on an explanation of the origin of variation, as was clearly recognised by Darwin (1868, e.g. pp. 430–431) in *The Variation of Animals and Plants Under Domestication* and not his contemporary critics (see also Bowler 1983, pp. 20–26). Nonetheless, evolution by natural selection does require that the process that generates variation does not determine the direction of evolution, which is instead set by natural selection on the generated variation. Accordingly, harmful traits could represent newly arisen variation that has yet to be eliminated by natural selection. This was the only explanation that was somewhat implicitly offered by Darwin (1868, p. 249), who remarked: 'though variability is indispensably necessary, yet, when we look at some highly complex and excellently adapted organism, variability sinks to a quite subordinate position in importance in comparison with [cumulative natural] selection'. Puzzlingly, in the discussion of hereditary diseases of the human eye, which Darwin (1859, p. 186) had previously cast as an organ of 'extreme perfection and complication', Darwin (1868, pp. 8–11) clearly recognises the ubiquity of hereditary diseases across species, their persistence down many generations, and their debilitating effects on individuals' ability to survive and reproduce, but does not explicitly reconcile their occurrence with his theory of evolution by natural selection. Instead, Darwin (1868, p. 11) rather weakly concludes: 'Seeing how hereditary evil qualities are, it is fortunate that good health, vigour, and longevity are equally inherited'.

This only further begs the question of why natural selection would cause the evolution of 'extreme perfection' of some traits alongside the obvious imperfection of others.

If not any sort of inefficiency of evolution by natural selection, why else might harmful traits exist? To take a specific example, there is an elevated frequency of a variety of hereditary diseases in the now millions-strong Afrikaner population of South Africa in contrast to the source populations in Western Europe. Type I progressive heart block is one such hereditary disease caused by irregular electrical signalling in the heart, which significantly increases the risk of cardiac arrest and other heart problems. The pedigree of this hereditary disease in the Afrikaner population has been mapped, which reveals its apparent origin in a par-ticular migrant couple—and its harmful consequences as the cause of ten related deaths in the first three generations of their children (Botha and Beighton 1983). Today, type I progressive heart block affects thousands of individuals in the Afrikaner communities of South Africa. So, is it right to imply that the hereditary disease is favoured by natural selection?

Not necessarily. Although the spread of type I progressive heart block can be mapped through the generations, its increasing frequency alone would not be enough to infer a role for natural selection because it is not the only evolutionary pressure that acts on populations. As well as natu-ral selection, there are other stochastic pressures that introduce a degree of randomness into the direction of evolution. This was certainly part of Darwin's world-view, but was somewhat neglected by his interpreters, including, influentially, the co-discoverer of natural selection Alfred Wal-lace. Despite fatefully coining the term 'Darwinism', Wallace's (1889, p. 103) own thinking went far beyond his co-discoverer's to suggest that natural selection's 'primary effect will, clearly, be to keep each species in the most perfect health and vigour, with every part of its organisation in full harmony with the conditions of its existence', which implies an all-powerful role for selection in shaping every trait (Cronin 1991, pp. 82–87). Contrasting against such use, Darwin's meaning of natural selec-tion was not a description of how a trait came to be 'just so', but an explanation of why one trait came to be rather than another (Gould and

Lewontin 1979). Accordingly, selection pressure is not about who survives per se because this may involve chance, but about how traits afford differential probabilities of survival, which is the hidden cause behind the regularities in who survives (Fisher 1930, p. vii). When considering the case of type I progressive heart block in the Afrikaner population, the disease clearly carries a substantial cost to individual survival through elevating the risk of cardiac arrest and other heart problems, so natural selection can be inferred to act against this trait. As such, the frequency of the trait is surprisingly high due to chance, which is reasonable given the detective work that established the migration of an individual with type I progressive heart block into the Afrikaner population, and the subsequent 13-fold increase in the size of the Afrikaner population within the first 100 years of colonial settlement (Van Der Merwe et al. 1994). Therefore, there is little reason to suspect that type I progressive heart block has persisted by natural selection for its own merits; instead, its frequency owes much to serendipity.

For similar reasons, most harmful traits have not been viewed as a great problem for evolutionary theory to explain because, as has been widely acknowledged for a long time, 'the explanation of this phenomenon must lie with the concept of the balance of evolutionary forces' (Gartler 1955, p. 40). But, in most cases, the evidence is not as compelling as for type I progressive heart block, and so the argument rests on the rather dubious practice of induction. To extend from a few known examples to a general theory was, admittedly, a key part of the Baconian scientific method (Popper 1963, pp. 12–16). As an inversion of such logic, the strength of the modern scientific method arises with tests that set out to falsify the deductions from theories. As Karl Popper (1972, pp. 15–18) famously elucidated, 'By this method of elimination, we may hit upon a true theory. But in no case can the method establish its truth', because failing to falsify a theory 'says nothing whatever about future performance, or about the "reliability" of a theory'. So whilst it is reasonable to admit the role of chance in the proliferation of some harmful traits, too little is known about the vast majority to say one way or the other. As the case in point, for the Afrikaner population, scientific knowledge of type I progressive heart-block owes particular thanks to their communities' dedication to its church records over the centuries. But such records stand in clear contrast to the knowledge of other human populations, let alone other species. In response to an absence of evidence, science differs from other areas of knowledge in its restraint: if the evidence is not available then the answer is left as unknown. Building on Popper's perspective, Peter Medawar (1967, p. 97) famously observed: 'No scientist is admired for failing in the attempt to solve problems that lie beyond [their] competence', because science is 'the art of the soluble'. Moreover, there is nothing to be directly gained from speculating about a question where evidence will never shed light one way or another because 'The method of science is the method of bold conjectures and ingenious and severe attempts to refute them' (Popper 1972, p. 81; see also Medawar 1967, p. 155). In this way, the puzzles posed by all harmful traits cannot be legitimately pinned to unidentifiable chance events rather than natural selection. This is particularly pertinent because harmful traits like hereditary diseases appear to be remarkably common across the human species; perhaps human populations are simply better studied, which is undoubtedly true, or maybe, for some of them at least, there is more going on than first meets the eye.

A clear example where there is more than reasonable doubt that chance is responsible for the persistence of a human trait comes from a hereditary disease of the eye itself. Leber's hereditary optic neuropathy (LHON) is a degenerative disease leading to the loss of vision. A study of LHON in Quebec found that a migrant woman brought LHON to the French Canadian population in the 17th century (Laberge et al. 2005). In their progeny, as is typical for LHON, males were eight times more likely to develop LHON than females (Milot et al. 2017). Further, it could be established that the hereditary disease had slightly increased in frequency at a rate of 3% per generation, with most of this increase between 1670 and 1920, before the introduction of modern healthcare practices. Furthermore, it could also be established that, on account of its apparent effects on males, including elevated infant mortality, lower marriage probability, and fewer children within marriage, the disease could have expected to strongly decrease at a rate of 36% per generation. As scientists have now established, something is causing this hereditary disease to persist even though it causes harm to (especially male) individuals—and, indeed, it is currently thought that LHON is likely to have persisted by natural selection for its own merits (Milot et al. 2017) for fascinating reasons that are explored in later chapters.

Harmful traits like hereditary diseases are a well-known phenomenon outside of scientific circles because of their far-reaching impact on human lives, but there are other examples from nature that are even more extreme for other species. These examples sit alongside a number of traits across the diversity of life that, at the outset, present a major problem for evolutionary theory to explain because of their surprisingly harmful effects on individuals' fitness. Why do worker ants sometimes kill their sister-queens that return to the colony after mating? Why is male infertility so common in populations of fruit flies? Why do many pregnant women suffer preeclampsia, putting their own and their babies' lives at risk? Why are many animals predisposed to cancer produced from their own cells? Why are animal, plant, and fungal genomes many thousands of times bigger than they need to be, with a large quantity of apparently useless 'junk' DNA? Amongst these traits, some of them stick out to the imagination as being obviously harmful, but all too often these traits have been normalised such that they fail to attract curiosity, masquerading as constraints on perfection that are mere 'facts of life'. Yet, for all of these examples and others, sufficient evidence has accumulated for there to be reasonable doubt that their persistence arises from the balance of natural selection and other evolutionary forces. Instead, as this book explores, there is a radical alternative explanation for some apparently harmful traits through the action of natural selection.

* * *

Evolutionary biology is a mature science with a long and successful history of understanding how natural selection has shaped the living world around us. Although hereditary diseases and similar traits do not clearly provide an advantage to individuals, much of recent evolutionary biology starts with the puzzle of apparently harmful traits and explains how they are really adaptations. The term 'adaptation' has a long-standing meaning going back to the Ancients (Amundson 1996), but the modern use of the term in evolutionary biology has its roots in William Paley's (1802) *Natural Theology* through its influence on Darwin (Desmond and Moore 1991, pp. 84–91). Paley argued that the obvious design of the traits of living things should be understood as adaptations for individual well-being with appeal to an external cause (in his case, a benevolent God). Such an explanation stands in contrast with alternative explanations that relied on the internal cause of will, which have been commonly grouped together under the popular name of 'Lamarckism', although (much like Darwinism) this term has come to mean something well beyond the *Zoological Philosophy* (1809) of Jean-Baptiste Lamarck (Corsi 2011). Paley's (1802, p. 34) reasoning against an internal cause like will was, quite simply, that a parent 'is in total ignorance why that which is produced took its present form rather than any other', so it cannot be responsible for the design of traits. Indeed, in a return to Paley's reasoning, rather than seriously entertaining Lamarckism as a competing scientific theory like Julian Huxley (1942, pp. 414–421) was compelled to in *The Modern Synthesis*, more current arguments have dismissed the role of the will in evolution as incapable of explaining how its own preferences for producing adaptations would arise in the first place (Cronin 1991, pp. 40–45; see also Madgwick 2021).

In contrast to Lamarckism, Darwin's theory of evolution by natural selection naturalised Paley's argument, switching the external cause of adaptation from a supernatural Creator to the external cause of the natural environment. In doing so, Darwin shifted the 'goal' of adaptations from the utility of contributing to individual well-being to the ability of individuals to survive and reproduce, whilst maintaining the logical structure of the argument with a creative designer and their created designs. As Stephen Jay Gould (2002, p. 158) reflected: 'one must buy into an entire conceptual world—a world where externalities direct, and internalities supply raw material but impose no serious constraint upon change; a world where the functional impetus for change comes first and the structural alteration of form can only follow'. Accordingly, despite the shift in goal, a specific trait like thicker fur remains an adaptation to the cold climate to reduce heat loss for Darwin as much as Paley. In transplanting the argument from theology to science, Darwin placed the problem of explaining design as adaptation at the centre of evolutionary biology: 'to explain the same set of facts that Paley used as evidence of a Creator' (Maynard Smith 1958, p. 82).

For as much as the design of a trait is apparent to the senses and with Darwin supplying us with the goal of design in contributing to individuals' abilities to survive and reproduce, adaptation is a conjecture that the 'means' of how the trait functions fits the 'end' of its design. Adaptation can be loosely inferred as a hypothesis for testing, as a counterfactual statement that the present design is more efficient than a plausible alternative at enabling an individual to survive and reproduce through some identified function (Williams 1966, p. 10). But to be accepted, evidence needs to demonstrate the closeness of fit between a trait's function and an aspect of the environment. As it is widely understood that 'Natural selection would produce or maintain adaptation as a matter of definition' (Williams 1966, p. 25), what is really being tested by acquiring evidence is the overriding evolutionary pressure from natural selection (rather than, say, chance events). The method of obtaining evidence is usually conducted through the comparison of trait efficiencies within different environments using experimental manipulation, or through contrasting closely related, ecologically similar, or geographically co-occurring species.

To demonstrate this method, a distinctively Darwinian illustration can be supplied. Paley's focus on well-being meant that his ascription of adaptation tended to focus on utilitarian traits that contribute to survival. But Darwin (e.g. 1871, pp. 214–250) clearly recognised that life poses many and varied challenges that can lead to surprisingly non-utilitarian traits that are adaptations for successful reproduction, such as the famous example of a peacock's train. A similar but neater example comes from a study of the long-tailed widowbirds *Euplectes progne* through an ingenious experiment (Andersson 1982). The males of this species stand out among other males from related *Euplectes* species for being jet-black, and producing extended tail feathers during the breeding season that are up to half a metre long (which is four times longer than their wingspan). In contrast, the female is camouflaged with a mottled-brown colouration and is otherwise inconspicuous in having tail feathers that are nearly eight times shorter than those of males. Both males and females need to have efficient flight for feeding and escaping predators, which may partly explain why males drop their tail feathers to look more like females outside of the breeding season. Why do males have such different tail feathers during breeding?

In the breeding season, the challenges of life for males differ from those of females to include holding territory against other males, attracting females to their territory using a vertical flight display, and persuading any and all passing females to mate with them. To test competing ideas about the benefits of their tail feathers, males can have their tail-feathers experimentally shortened, elongated, or kept the same, to compare the breeding success of males to determine what tail feathers might be an adaptation for (Andersson 1982). Experimental evidence shows that changing tail-feather size does not impact the ability of males to hold territory or alter the size of their territory, though males with shortened tails are slightly more active in territory defence and more frequently engaged in the vertical flight display (presumably, from being unencumbered by the weightier tail feathers). The longer tail feathers make a male more visible during the vertical flight display, where the male slowly rises up to two metres into the air and slowly falls back towards the ground, with the unaltered tail feathers making the lateral outline of the jet-black male around three times larger than a male with shortened tail feathers. Yet, males display relatively close together, which means that females are easily able to see a male at a distance irrespective of their tail-feather length. Indeed, females tend to visit multiple neighbouring males in quick succession, but they only mate with the males that have the longest tail feathers. Therefore, the key to understanding the male widowbird's tail feathers rests in mating success being lowest for males with shortened tails and highest for males with elongated tails, which suggests that longer tail feathers are a reproductive adaptation to suit a female mating preference: longer tails make a male more attractive to a female widowbird. Adaptations like this can be surprisingly non-utilitarian because the environment that generates natural selection encompasses a varied range of demands on living things. Nonetheless such traits still make an individual more efficient at the complex task of surviving and reproducing.

Some traits do not have an obvious functional benefit to individuals through survival or reproduction, like predispositions for infertility, eclampsia, or cancer. If a trait does not admit explanation as an adaptation through its obvious or cryptic efficiency, does that make it a maladaptation? In evolutionary biology, the term 'maladaptation' has no widely accepted definition to furnish it with a precise meaning, and instead it has loosely been applied to a broad range of phenomena. Richard Dawkins (1982, pp. 48–53) argues that what others have referred to as maladaptive traits should really be thought of as 'adaptations under constraint', but he does recognise that some environments are unlikely to lead to adaptation. Bernard Crespi (2000) goes further to define maladaptation as any deviation from adaptation, arguing that there are wide-ranging aspects of genetic evolution that can cause this to occur, including both systematic factors like a preference for inbreeding, and chance factors like a particular individual migrating to a new environment where it struggles to live. In keeping with the same theme but in the context of the relationship between evolution and disease, Randolph Nesse (2005) defines maladaptation by example, applying the term loosely to cases where the environment has changed or is changing faster than selection, where there are physical constraints that limit the efficiency of traits, and where a defensive action by the immune system is mistaken for a disease.

The themes in the definitions of Dawkins (1982), Crespi (2000), and Nesse (2005) run through to the present (Brady et al. 2019b), where the term maladaptation is still used in application to a wide range of phenomena without clear unifying features. Such inclusiveness is unsurprising when it is recognised that maladaptation is a neglected concept. Indeed, whilst the term 'adaptation' has been used thousands of times per year in scientific publications in the areas of evolution and ecology over the past two decades, 'maladaptation' is used but tens of times—mostly in the context of anthropogenic extinction (Brady et al. 2019a). Beyond just being inclusive, there is also conflict between definitions, which can become confusing in the conflation of traits that are maladaptive because of different evolutionary pressures. For instance, inbreeding may cause maladaptation in the sense that it weakens the selection for adaptation, a changeable environment may cause maladaptation in the sense that there is a need to reassess what selection really favours (i.e. what constitutes adaptation in this context), and the chance migration of an individual to a new area may have nothing whatsoever to do with natural selection. Such fuzziness about the causes of maladaptation stands in stark contrast to the relatively strict usage of the term adaptation to apply to a trait that is the result of the natural selection for a function in response to a specific aspect of the environment.

Whilst previous discussions of maladaptation have usefully expanded the horizons of evolutionary biology to recognise that the natural world is not only full of adaptations, those discussions also have their limitations. As current perspectives on maladaptation entirely neglect, there are cases where natural selection really does favour a trait that harms individuals. Or, more exactly, natural selection favours a trait *because* it harms individuals, not *despite* the harm it causes. These traits surely deserve the title of 'maladaptation' for literally being 'bad' or 'evil' adaptations, whilst other traits that are not adaptations for other reasons deserve some other description. As such, doing away with a fuzzy conception of the term, maladaptation can be given a fresh definition, as a consequence of the action of natural selection.

Maladaptation will be used to refer to an instance where natural selection acts in what would classically be viewed as the 'wrong direction' by favouring a trait in an individual that decreases the ability of that individual to survive and reproduce. This is a highly restrictive definition of maladaptation, which excludes instances of what will be referred to as non-adaptation where natural selection does not occur or is not involved, such as resulting from chance events that limit the beneficial variation of heritable traits or where the environment changes faster than organisms become adapted. Whilst it is conceded that the causes of non-adaptation are often complicated in arising from the interactions of various chance events, the causes of maladaptation, as has been defined, are just as explicable as those of adaptation because both only arise because of natural selection. Herein is exposed the rationale behind the concept: the proposed definition of maladaptation has an intuitive symmetry with the existing concept of adaptation, rather than inclusively describing any deviation from adaptation.

As consequence of the symmetry, many of the examples of maladaptation that will be discussed come from complex traits that have previously been described as adaptations. This is important because maladaptations are not 'broken' traits, as they can be equally as fine-tuned by natural selection. For example, pleometrosis occurs when new ant queens found a colony together (Wilson and Hölldobler 1990, pp. 217–220). The queens cooperate in the early stages of colony formation, but struggle against each other for sole control of the mature colony that will help them to produce reproductive offspring. Estimates of the benefits of cooperative colony foundation commonly find a concave (i.e. an ∩-shaped) relationship between fitness proxies and the number of foundresses. For instance, in the honeypot ant *Myrmecocystus mimicus*, the number of eggs, larvae, pupae, and workers is optimal with three foundresses (Bartz and Hölldobler 1982). Even ignoring that there are often more foundresses in nature, three cooperative foundresses tend to produce fewer workers per queen in the early stages of colony growth compared to solitary foundresses because, as is common across species (Bernasconi and Strassmann 1999), queens eat each others' eggs. Further, given that only one queen will eventually reproduce, it seems likely— and consistent with current evidence (Crozier and Pamilo 1996, pp. 123–135)—that cooperative queens on average produce fewer offspring than solitary queens. Definite evidence is, unfortunately, lacking for reasons that will be discussed in later chapters. But, in the wider context of polygynous ant colonies with more than one queen, the suggestion is highly plausible, especially in the more extreme cases like *Formica uralensis* that have over five hundred queens per nest; whereas what are widely regarded as the most massive and complex colonies of *Atta* leafcutter ants succeed with only a single queen from colony foundation (Wilson and Hölldobler 1990, pp. 209 and 596). A hypothesis of pleometrosis as a maladaptation in these cases would suggest that these complex phenomena are not cryptic adaptations as current working hypotheses assume (Crozier and Pamilo 1996, pp. 123–124), but rather represent natural selection towards a qualitatively different kind of trait that is harmful to fitness.

A more rigorous understanding of the meaning of maladaptation will be developed in the book, but for now it is important to recognise that some evolutionary biologists would not agree that maladaptations that harm individual survival and reproduction can—or even could—be favoured by natural selection. Indeed, this kind of maladaptation does not enter into current definitions or discussions of maladaptations (e.g. Brady et al. 2019a). Consequently, it would be misleading to propose that maladaptations be cast as a subcategory of adaptation alongside something like 'beneadaptations' as good-adaptations, which some might be inclined to prefer, because the current meaning of adaptation always refers to its beneficial case. The explanation for this goes all the way back to foundations of evolutionary biology. Paley (1785, pp. 40–41), in developing his natural theology, established that there was a benevolent designer:

> The world abounds with contrivances: and all the contrivances which we are acquainted with, are directed to beneficial purposes. Evil, no doubt, exists; but is never, that we can perceive, the object of contrivance. Teeth are contrived to eat, not to ache; their aching now and then is incidental to the contrivance, perhaps inseparable from it: or even, if you will, let it be called a defect in the contrivance: but it is not the object of it. . . .. We never discover a train of contrivance to bring about an evil purpose. No anatomist ever discovered a system of organization calculated to produce pain and disease.

Darwin (1859, p. 211), by substituting the environment into the role of the designer, transferred the benevolence of God that is expressed in design to natural selection—and even referenced Paley as he did this:

> Natural selection will never produce in a being any structure more injurious than beneficial to that being, for natural selection acts solely by and for the good of each. No organ will be formed, as Paley has remarked, for the purpose of causing pain or for doing an injury to its possessor. If a fair balance be struck between the good and evil caused by each part, each will be found on the whole advantageous.

Darwin is not making a peripheral or throw-away comment; it is repeated throughout *The Origin of Species* (pp.109, 125, 172, 233, 471, etc.) as a cornerstone of his explanation of how individuals acquire traits. For the vast majority of evolutionary biologists since Darwin, natural selection is implicitly benevolent in this Paleyan sense. By challenging this fundamental maxim of evolutionary biology, the claimed existence of maladaptation is likely to be received as a radical proposition by some evolutionary biologists, let alone other biologists, scientists, and the general public.

To understand just how radical it is, consider its most extreme implication. As was previously a source of some controversy in the argument for group selection (*sensu* Wynne-Edwards 1962), evolutionary biologists now well accept that natural selection, in furnishing individuals with traits that help them to survive and reproduce, does not always help groups of individuals or act for the good of the species (Dawkins 1976, pp. 7–8). But group selection is an innocuous theory compared to maladaptation because it at least relies upon natural selection on individuals and groups acting in the same direction. By contrast, at its most extreme, maladaptation could state the complete opposite: that natural selection alone could drive a species to extinction by reducing the ability of individuals to survive and reproduce. As later chapters explore, it is not possible to draw a universal or straight-forward link between a reduced ability to survive and reproduce and population extinction—so it is important not to trivialise their relationship. Nonetheless, even this possibility would certainly have baffled Paley and Darwin.

The potential for maladaptation to lower fitness and hence drive extinction would also have puzzled some of the biggest names from the diverse traditions of modern evolutionary biology, like Huxley, Gould, Williams, and Maynard Smith—to pick but a few that have already been mentioned. Huxley (1942, pp. 439–440), as a staunch Darwinian individualist, echoes the view that 'a balance will eventually be struck at which the favourable effects slightly outweigh the unfavourable'; whilst he suggests that any 'intraspecific competition is on the whole a biological evil', he means this in the irrational sense that 'Natural selection, in fact, though like the mills of God in grinding slowly and grinding small, has few other attributes that a civilized religion would call Divine'. Similarly, but from a multi-level selection perspective, Gould (2002, p. 127) argues that

> selection works directly for the benefit of organisms only, and not for any larger harmony that might embody God's benevolent intent. Ironically, through the action of Adam Smith's invisible hand, such 'higher harmony' may arise as an epiphenomenal result of a process with apparently opposite import—the struggle of individuals for personal success.

So neither Huxley nor Gould thought that natural selection reliably produces the best of outcomes for individuals, but it was in no way deliberately malevolent towards them. Williams (1966, pp. 26–28), taking a gene's eye view, agreed that 'We ordinarily expect selection to produce only "favourable" characters', and that a 'frequent outcome of natural selection is the promotion of the long-term survival of the population'; but, again like Darwin (and Huxley), he did admit exceptions where harmful effects are a by-product of adaptation (assuming that the helpful effects outweigh it). These exceptions include examples like how 'evolutionary increases in body size [can] cause a decrease in numbers, and this might contribute to extinction' (Williams 1966, p. 28). But they are unlike maladaptation because, in this case and others, it is merely that smaller populations are at greater risk of extinction, rather than natural selection favouring a trait that actually lowers fitness to drive extinction in of itself. Indeed, a similar absence of such reasoning exists in the pragmatic thinking of Maynard Smith, who excludes anything that could be described as natural selection driving fitness decrease or population loss in an influential discussion of the causes of extinction (Maynard Smith 1989).

Other more contemporary evolutionary biologists that study some touchpoints of maladaptation may well already be familiar with some potential examples, including from natural populations. For instance, the common eggfly *Hypolimnas bolina* is a tropical butterfly with populations on many Pacific islands, including the Samoan islands of Upolu and Savaii. It is well known in evolutionary theory that natural selection often favours an equal sex ratio with individuals producing as many female offspring as male (Fisher 1930, pp. 141–143), which is known to be true of the common eggfly based on long-running field sampling and laboratory rearing (Hornett et al. 2009). Yet, in 2001, the sex ratio of the population reached >99% female because females almost exclusively produced female offspring (Charlat et al. 2007), which did not occur by chance but as a heritable trait. Such bias did not result in population extinction in the Samoan cases, based on the return to a near-equal sex ratio by 2006. However, in Fiji, museum specimens suggest that the local population did go extinct between 1886 and 1934 due to a similar fluctuation in sex ratio (Hornett et al. 2009). As such, whilst maladaptation is a bold conjecture, there is support from as-of-yet unsynthesised evidence. The scientific challenge here is to bring such examples together to make the first steps towards a systematic theory of maladaptation to explain when and how it can occur.

* * *

The first aim of this book is to make the case that maladaptations that reduce individual survival and reproduction are theoretically possible. Chapter 2 explores how natural selection on individuals, as envisaged by Darwin, differs from natural selection through genetics, as modern theory describes. Starting with the discoveries of transmission genetics, the history of evolutionary biology is discussed through to Huxley (1942) heralding the unification of Mendelian genetics and Darwinian natural selection into a coherent theory of evolution in *The Modern Synthesis*.

Following the thread of evolutionary genetics going forward, the development of the gene's-eye view of evolution is explored, as championed by Dawkins (1976) in *The Selfish Gene*, wherein the understanding of natural selection was radically transformed to become fundamentally gene-centred. Finally, building on the implications of molecular genetics from Helena Cronin's (1991) argument in *The Ant and the Peacock*, it is argued that a functional understanding of a gene is necessary to understand what is really under selection, which is essential to the understanding of how maladaptation can evolve by natural selection.

Chapter 3 goes further to present maladaptation as an overlooked possibility because it requires the integration of population ecology into evolutionary theory. Maladaptation arises from the disharmony between natural selection favouring individual fitness and gene replication, which amounts to a distinction between absolute and relative success. The logic of a competition favouring absolute or relative success is established using sports, before turning to consider the specific details of competition among alleles. Having established its possibility, the reasons for it being overlooked are then examined, linking back to the foundations of evolutionary biology in the ideas of Paley and Darwin, which was formalised into evolutionary theory by Ronald Fisher. Whilst it is possible for the underlying error of reasoning to be patched up by the reframing of evolutionary theory, it is argued that there is a need for the concept of maladaptation in evolutionary biology for the same reason that a modern economist needs the concept of 'market failure': to understand that even though the prevailing direction of change may bring about a benevolent outcome in idealised theory, there are predictable reasons that regularly bring about malevolent outcomes in reality.

The second aim of this book is to find evidence of maladaptation. As becomes apparent, it is often difficult to find definitive evidence. The purpose here is not to create a complete list of maladaptations, which would be of as much use as a comparable list of adaptations. Instead, the purpose is to use examples to showcase some of the most important theoretical considerations when seeking evidence of maladaptation in the real world. In Chapter 4, maladaptation is sought among examples of social behaviour. Three pitfalls are identified in assuming that maladaptation is synonymous with examples of intraspecific competition, spite, and/or genetic conflict. The following are discussed: how intraspecific competition could be beneficial in allocating resources to those that can use them best, how spite is classified ambivalently because inclusive fitness is designed to predict the direction of selection, and how genetic conflict describes a potential that may not actually harm individual fitness.

Chapter 5 extends the search for maladaptation to traits within the body of an individual organism, which may not have the most obvious examples, but it is where current evidence arguably supports the most convincing cases of maladaptation. Numerous examples of meiotic drive are explored before turning to consider selfish genetic elements more widely—especially in some of the key examples and conclusions of the long review of Austin Burt's and Robert Trivers's (2006) *Genes in Conflict*. The widely held view that selfish genetic elements are unimportant to organism design is rejected, along with the associated egalitarian theory of the 'parliament of the genes' (Leigh 1971, p. 249). Suppressors are often assumed to be all-powerful, but are argued to be weaker at suppressing maladaptations, especially when they are pleiotropically constrained, permitting selfish genetic elements to have a large impact on organism design. Consequently Chapters 4 and 5 conclude by recognising the need to shift to a non-egalitarian theory of the 'society of genes' (Yanai and Lercher 2016, pp. 43–46), where it is accepted that genes do not need to reap the same benefit to cooperate with one another in the production of organismal traits.

Beyond their being a reality, the third aim of this book is to make the case that traits that reduce individual survival and reproduction are not an 'odd bunch' of strange phenomena that arise in unusual circumstances, but instead are a part of the architecture of the natural world down to its very foundations. Whilst Chapters 4 and 5 worked through specific examples of maladaptation, Chapters 6 and 7 delve into greater speculation in the service of building a broader concept of maladaptation. Chapter 6 draws upon John Maynard Smith and Eörs Szathmáry's (1995) *The Major Transitions in Evolution* to link maladaptation in social behaviour with maladaptation in the body by examining how the biological entity that is referred to as 'the individual organism' has evolved in the history of life on Earth. Many traits of earthly life that contribute towards organismal complexity are argued to be underpinned by maladaptation because of a basic flaw: genes are used in the building blocks of new individualities, despite those genes retaining their evolutionary potential.

Chapter 7 develops this idea further to ask whether all life anywhere in the universe would exhibit maladaptation by discussing the most basic living things in the deep history of life. Following Iris Fry's (1999) distinction in *The Emergence of Life on Earth*, ideas on the origin of life are broken into replicator-first and organism-first theories, and exploring their hybrid forms, which emphasises the separate evolutionary potentials of genes and individuals. Building on the ideas in Graham Cairns-Smith's (1982) *Genetic Takeover*, a thought experiment of chemical evolution is used to illustrate a tendency for a process akin to evolution by natural selection to favour chemicals based on relative success, which must be overcome for the dynamic system to avoid rapid extinction. A speculative explanation is offered that identifies the long-standing association between life and maladaptation. On the basis of its role in the origin of life and all that follows, maladaptation is argued to be integral to the design of all life anywhere in the universe.

With the three overarching aims of showing the possibility, reality, and importance of maladaptation, the case that this book builds should not be misconstrued as anti-Darwinian. There is nothing 'false' about the explanation of the appearance of design in adaptations through evolution by natural selection. Instead, maladaptation offers a complement to the explanation of adaptations by explaining why selection can lead to other outcomes. As such, the identification of an unambiguous example of maladaptation would not reveal some great anomaly in current evolutionary theory that could provoke a 'paradigm shift' to a new theory of evolution (*sensu* Kuhn 1962), which (anyway) is simply not how the life sciences work. Paley (1802, p. 77) got this right when he remarked:

proof is not a conclusion which lies at the end of a chain of reasoning, of which chain each instance of contrivance is only a link, and of which, if one link fail, the whole falls; but it is an argument separately supplied by every separate example. An error in stating an example, affects only that example. The argument is cumulative, in the fullest sense of that term.

So too each putative case of maladaptation is a separate source of evidence, but whilst some of the examples reflect traits that are peculiar to particular species, others are more widely shared across the diversity of life on Earth—and maybe even beyond Earth. In this way, building on the research tradition that stems from Darwin, the argument advocates that room must be made for maladaptation in the understanding of nature, as a fundamental yet all-too-often overlooked aspect of the design of individual organisms across the diversity of life.

The final chapter explores what the existence of maladaptation might mean for us. Returning to the question posed at the start of this chapter, it is asked: why do living things appear designed, and what are they designed for? The scientific response to this question is vague, which may fall short of a satisfactory answer because the question is not an exclusively scientific one. Its philosophical and theological intimations are discussed, provocatively asking: if Paley used adaptation as evidence of the benevolence of the Creator, should maladaptation be used as evidence of their malevolence? Delving into Paley's reasoning further, which is much more sophisticated than is usually given credit, it is established that the scientific claim of maladaptation would have presented a special problem to his argument. Nonetheless, building on the spirit of Paley's reasoning, despite Darwin's own beliefs and those of many evolutionary biologists since, it is argued the discovery of evolution by natural selection (and maladaptation) cannot be said to have demolished the design argument. Consequently, a pluralism on the meaning of design in nature is defended, in the hope of protecting the nascent study of maladaptation from its abuses outside of science.

Overall, then, whilst this is a book about science, it is also trying to do more than just present and organise facts: the goal is to inspire curiosity. Critics will no doubt find weaknesses in the case that is presented because, as with any new idea, there is almost always going to be ambiguous support from existing evidence. Gone are the days where new theories are published only with sufficient supporting data, in part because science is no longer the pursuit of a few bored gentlemen finding something to do in their leisure, as it largely was at the time of Darwin. Instead, science is now a professional industry, where new ideas provide the fuel for future research. Some scientific works, like this one, are necessarily more 'entrepreneurial' than others in carrying greater risk in boldly presenting a new perspective in the hope that the resulting insights can enable new discoveries. As such, whilst readers are asked to open-mindedly exercise their judgement, if support is only ascertained when the scientific evidence is incontrovertible then an opportunity has been missed for presciently discerning the truth ahead of the evidence by the mastery of theory. On this point, the perspective put forward in *The Selfish Gene* springs to mind; for all that the book is lauded for being 'astonishing [in] how lucid and absolutely right all the detailed arguments are' (Grafen 2006a, p. 73), it is debateable whether we would still be talking about it fifty years later were it not for the subsequent discovery of selfish genetic elements that demonstrate the core premise by only functioning to serve their own replication. Similarly, perhaps the eventual discovery of a natural or engineered maladaptation that drives a population to extinction could demonstrate that the argument puts forward more than a perspective to pinpoint an overlooked truth about reality. Until that point, the overarching aim of this book is to inspire curiosity with enough confidence to commit to those first steps on the road to discovery.

2

Natural selection through genetics

Before beginning to understand maladaptation, a more sophisticated understanding of the evolutionary process is needed than was known about and presented by Darwin. The difficulty for Darwin's argument for evolution by natural selection is that it doesn't really tell us what is selected (Alexander 1979, p. 23)—and this cuts to the heart of the difficulties surrounding the early studies of natural selection. For instance, Bernard Kettlewell's (1958) analysis of melanism in the peppered moth *Biston betularia* is held up as a classic example of natural selection in action, where the sooty environment from industrial revolution in Britain gave dark-coloured moths a camouflage advantage against predating birds over their light-coloured counterparts. This change can be documented in the changing frequency of dark and light colourations over many years, and also in comparison between cleaner rural and sootier urban settings. Further, when the Clean Air Act (1956) was introduced, the cleaner city environment meant that many urban moth populations saw the frequency of light-coloured peppered moths come to dominate populations once again (Majerus 1998). With all this evidence, it might seem reasonable to concur that the dark-coloured peppered moths were favoured by natural selection in sooty urban environments, but it must be remembered that a moth does not directly pass on its colouration to future generations. Instead, there is some hidden mechanism whereby parents pass on their traits to their offspring—as like tends to beget like.

It is reasonable to think that since Kettlewell showed a changing trait in response to a changing environment, the evidence of natural selection's involvement is convincing, but this ignores the other possible scenarios where a trait that is favoured by natural selection does not come to dominate in the population. A classic counterbalancing example here is Anthony Allison's (1954) analysis of protection against malaria in human populations across the tropics. Malaria is sometimes deadly but more often debilitating, as an infectious disease caused by the parasite *Plasmodium* that transmits between humans through mosquito bites. Accordingly, protection against malaria might be expected to be strongly favoured by natural selection, but there are complications. Individuals with protection against malaria are more likely to be healthy adults that go on to be parents. Yet, their children are often affected by sickle-cell anaemia, a heritable disease affecting beta-haemoglobin that leads to red blood cell malformation, causing a range of symptoms from painful episodes to increased risk of other infections (that can also be deadly). Seemingly by consequence, protection against malaria is found at intermediate frequencies within human populations in the tropics. So even though parents with protection may be better at surviving and reproducing in environments with mosquitos, the trait of 'protection against malaria' does not spread throughout these human populations. Herein, exactly what is passed between the generations is of critical importance to whether or not the protection against malaria increases in frequency by natural selection, even though it is favoured by natural selection, because protection against malaria in parents is associated with sickle-cell anaemia in their children. Therefore, comparing Kettlewell's and Allison's results, there is no general basis for understanding why natural selection did what was expected for dark colourations in peppered moths and not what was expected for protection against malaria in humans. What is going on here?

The comparison between Kettlewell and Allison's studies shows that, to understand how evolution works, it is necessary to understand the inheritance of traits—at least, in basic terms. In this topic, it is highly useful to have a historical perspective because, to a large extent, understanding how ideas developed helps to explain the peculiar concepts that are used in evolutionary theory to the present day.

The logical structure of inheritance was established in the foundational work of Gregor Mendel, which was translated for the English-
speaking scientific community in 1902. For over a decade, Mendel (1866)
famously examined the pattern of inheritance of seven mutually exclusive
traits from flower colour (purple or white) to plant height (tall or short)
in the common garden pea. This was an exceptionally useful system
for the study of inheritance because garden peas can be self-pollinated;
repeated inbreeding enabled Mendel to generate lineages of individuals
where offspring always possess the all the traits of their parents. Such
'breeding true' is an interesting result in its own right, but does little to
illuminate the causal mechanisms at work behind the general principle
of like begetting like.

Mendel performed two simple crosses between individuals from different inbred lineages to describe the foundations of the genetic theory
of inheritance—although many of the terms to describe what was going
on were coined much later. First, Mendel crossed Fo parents from two
inbred lineages and observed that it was typical for their F1 offspring to
inherit the trait of just one parent irrespective of their sex; for example,
an individual with purple flowers crossed with an individual with white
flowers leads to all offspring having purple flowers. The result demonstrates that some traits have dominant inheritance whilst other have
recessive inheritance, which is not to do with preferential inheritance
from a particular sex (as the Ancients believed) but instead is a constant
and repeatable property of the trait itself. Second, Mendel crossed the F1
offspring together and observed that it was typical for there to be three
times as many F2 offspring with one trait like purple flowers over another
like white flowers. With this result, Mendel demonstrates that the trait-
causing qualities of what is inherited from each parent are unchanged by
the process of inheritance because a trait can skip generations to recur
after its absence, which runs counter to ideas about inseparable blending
in the process of inheritance. Instead, Mendel postulated a hypothetical
particle of inheritance, which is now known as a gene, that parents must
replicate into their offspring; after all, a parent's ability to produce purple flowers is not used up in the process of passing the trait on to their

offspring. Further, the quantitative three-to-one pattern of dominance suggested that individuals have two copies of every gene—one inherited from each parent.

Third, as is often neglected, Mendel observed that different mutually exclusive pairs of traits, like either flower colour or plant height, displayed the same pattern of inheritance irrespective of the parental combinations of traits. Accordingly, in a cross of one Fo parent that is tall with purple flowers with another Fo parent that is short with white flowers and then the subsequent cross of F1 offspring together, the result is indistinguishable from the cases where one Fo parent is short with purple flowers and the other Fo parent is tall with white flowers. Consequently, Mendel argued that the inheritance of different traits occurs independently, giving rise to the concepts of alleles and loci. An allele describes a functionally equivalent class of genes that are associated with a particular trait, like the allele for purple flowers being different from the allele for tall plants. A locus describes a conserved place for a gene within the heritable material of an individual, where each parent always passes on one half of the alleles that an offspring inherits—giving one allele per trait rather than randomly giving half of all their alleles across traits. So each individual has two alleles per trait—one from each parent—that reside at a locus in an individual's heritable material that houses variation in that trait, with the locus for flower colour having purple and white alleles that interreact to give an individual its flower colour trait.

The achievement of these experiments is extraordinary in providing the foundations for the study of inheritance in the science of genetics, but reality was much messier than Mendel's work implies. To some modern eyes, Mendel was either lucky in chancing upon species and traits with simpler modes of inheritance or performed a degree of 'cherry-picking' in only reporting results from those species and traits that fitted his theory. Further, it is clear that Mendel regarded the process of inheritance to be a random one, such that the regularities were probabilistic (not deterministic), but there was still a major problem in the missing explanation of variation. Indeed, there is something 'static' about Mendel's description of inheritance that misses out how the heritable variation may change

over the generations in a gradual process of evolution; this is most starkly
seen in that there was no explanation for the process that originates new
variation.

The history of genetics tends to think of Hugo de Vries (1901–1903)
as birthing the idea of mutation as a sudden change in the heritable material of an individual, but the philosophy of transmutationism has a long
history going back to the Ancients. Further, de Vries's views were a far
cry from the contemporary understanding of a mutation as a random
modification of a gene to create a new allele. The modern meaning seems
to have originated rather gradually, perhaps with the strongest influence
coming from Thomas Morgan, who identified a series of small mutations
that gave rise to new alleles for easily identifiable traits within laboratory
populations of the common fruit fly. Morgan et al. (1915) went on to
build on Mendel's insights with the first exposition of familiarly modern
genetics based on chromosomes. Despite Mendel's work suggesting that
traits are independently inherited, this is not always true. Indeed, Morgan
observed that some traits have an elevated probability of co-inheritance.
This led to postulation that genes were inherited on a number of chromosomes that represented physical strings of heritable material, which
alleles adorned like beads. In general, one copy of each chromosome is
inherited unchanged from each parent, but it was a matter of observation
that sometimes an individual would appear to deviate from the expected
pattern. Such events were referred to as cross-overs, imagining that por-
tions of the homologous pairs of chromosomes from each parent may
switch places. Morgan used the probability of co-inheritance to map alleles onto chromosomes in an almost trigonometric manner, relying upon
the observed pattern that if alleles A and B independently have a high
probability of co-inheritance with allele C, then it follows that allele A
has a high probability of co-inheritance with allele B. In this way, Morgan explained complicated patterns of co-inheritance through the simple
idea that particular alleles were physically linked together.

Amazingly, the important features of genetics for evolution were outlined based on the rudimentary insights of Mendel, Morgan, and others,

which was sufficient for the formulation of a mathematical theory of evolution by natural selection in population genetics, without needing to know anything about the physiochemical processes at work. As remains the case to the present day, the utility of the mathematical theory of evolution by natural selection was its power to interrogate ideas in evolutionary biology by rigorously sifting through the logic underpinning the competing views to establish which theories have self-consistency. At the start of the twentieth century, there were many competing views about the nature of inheritance that influenced the development of genetic theory, but the logic that Mendel established by experiment eventually became the bedrock of the new science of genetics because it was the only theory of inheritance that did not lead to the rapid loss of new variation within populations, which is necessary for the possibility that advantageous traits could spread through populations. Herein, drawing heavily upon their own independent contributions to population genetics, Fisher (1930), 'JBS' Haldane (1932), and Sewall Wright (1931) are celebrated as demonstrating the compatibility of Mendelian genetics through discrete particles of heritable material and Darwinian evolution by natural selection on individual organisms, which was famously championed as a successful unification affording mutual validation in Huxley's (1942) *The Modern Synthesis*.

So, what does transmission genetics add to the understanding of the difference between Kettlewell's (1958) and Allison's (1954) studies? It is a matter of observation that peppered moths are diploid, carrying two of every gene, one from each parent. Further, it has been established that dark colouration is dominant; the expression of this trait occurs when an individual carries a single gene of that allele. More complicatedly, in human populations, it has been established that protection against malaria is dominant, but it is linked to sickle-cell anaemia that is recessive, in being expressed when an individual carries two genes of its allele. This is the structure of the inheritance of these traits; it does not tell us how these traits are expected to change under evolution by natural selection.

* * *

As not everyone immediately recognised, including Huxley in *The Modern Synthesis*, genetics did not really vindicate Darwin's original argument for natural selection; instead it radically transformed it. As Peter Medawar reportedly said, 'The trouble with Julian [Huxley] is that he really doesn't *understand* evolution' (Dawkins 2013, p. 269, emphasis in original). In the minds of population geneticists, evolution was changing from the epic process of organism design over the millions of years of geological time, as described by Darwin, to an often-rapid process of changing traits on more human timescales, as observed in the new wave of examples like industrial melanism in peppered moths. Moreover, a subtler shift in the way in which evolutionary theory was discussed was also taking place, neatly summarised in the redefinition of evolution itself as 'a change in the genetic composition of populations' (Dobzhansky 1937, p. 11). Although Huxley tried to use genetics to vindicate Darwin's argument for evolution by natural selection after a period of its 'eclipse', Darwin's conception places individual organisms as the key players in evolutionary change, whereas genes were beginning to take their place in some quarters.

Of the early population geneticists, Fisher (1914, p. 315) perhaps most clearly saw that genetics radically transformed evolutionary biology when he wrote that if 'A is the eldest son, and stays at home; his brother B goes to the wars; then so long as A has some eight children, it does not matter, genetically, if B gets killed, or dies childless, there will be nephews to fill his place'. Later, Haldane (1955, p. 44) similarly mused:

> Let us suppose that you carry a rare gene which affects your behaviour so that you jump into a river and save a child, but you have one chance in ten of being drowned . . . .. If the child is your own child or your brother or sister, there is an even chance that the child will also have the gene, so five such genes will be saved in children for one lost in an adult. If you save a grandchild or nephew the advantage is only two and a half to one.

Both Darwin and the early population geneticists agreed that individuals are the bearers of traits that are subject to natural selection, but the population geneticists realised that the outcome of selection was about how those traits influence the transmission of genes between the generations. Accordingly, when thinking about evolution, an individual's 'value' to another individual comes from the number and identity of the genes it carries.

This perspective was crystallised into systematic theory by William Hamilton (1963; 1964a, b) in the addressing the problem of altruism, especially in the evolution of worker sterility in ants, bees, and wasps where a worker appears to forgo their own reproduction to help their queen to reproduce. The concept of altruism was influenced by Haldane (1932, pp. 207–210), who had previously discussed the general problem of altruism in the context of group selection in humans, which was topical at the time in centring on the altruistic act of fighting for one's country, where an altruist's relative disadvantage within a group is counteracted by the relative advantage in comparisons between groups. But when it works, this explanation essentially says that an apparently altruistic trait within a group is not really altruism because it net-benefits altruistic individuals, which is of no help to the explanation of worker sterility where an altruistic individual never reproduces. The problem of worker sterility was identified much earlier by Darwin (1859, p. 305), who admitted that 'this is by far the most serious special difficulty which my theory has encountered' because the adaptations of sterile workers could not be acquired through the differential survival and reproduction of sterile workers. Consequently, the obvious adaptations of workers pose a problem to the general theory of natural selection among individuals. Nonetheless, Darwin (1859, pp. 300–301) did provide a suggested explanation of sorts with appeal to domesticated animals:

> This difficulty, though appearing insuperable, is lessened, or, as I believe, disappears, when it is remembered that selection may be applied to the family, as well as to the individual, and may thus gain the desired end. Breeders of cattle wish the flesh and fat to be well marbled together: an animal thus characterised has been slaughtered, but the breeder has gone with confidence to the same stock and has succeeded.

This 'family effect' was used as the first explanation for altruism in worker sterility (Williams and Williams 1957), but it should not be misunderstood as anything other than an individual-centric description and, as such, there was a key missing piece.

Hamilton (1964a) asked a straightforward question: what property does natural selection act as if to maximise? Fisher (1930, pp. 34–37) supported Darwin's account that individuals have traits that act as if to maximise their fitness, measured as their 'total lifetime reproductive success' in the number of offspring that an individual has over the course of their life. Like Darwin (1859, pp. 300–301) with his 'family effect', Fisher (1930, pp. 160–162) recognised exceptions involving the 'propinquity of kin', including in the evolution of distastefulness in certain social insects, but (like Williams and Williams 1957) he did not identify a broader principle at work. Hamilton (1963; 1972) argued that natural selection acts on traits as if to increase an individual's contribution of their genes for that trait into the next generation. Hamilton (1964a) translated this argument to mean that natural selection acts as if to maximise each individual's 'inclusive fitness', which is their direct fitness in the number of offspring that they produce and, inclusively, their indirect fitness in the number of offspring that they help those that share genes to produce. The spectacular conversion of a statement about the selection of genes into an expectation about individuals' traits is performed by the term for relatedness. As Hamilton (1963) identified, indirect fitness depends on both the benefit donated to another in terms of the increased number of offspring produced, but also the relatedness between the recipient of the altruism and the actor, where relatedness describes the probability that the recipient shares the actor's gene for altruism. Hamilton (1964a) explored the general mechanism of sharing genes via kinship to explain worker altruism in social insects, where a gene is shared by relatives through descent from a common ancestor: in social insects, sterile workers help their mother to reproduce the next generation, which means that a sterile worker helps to produce reproductively capable siblings, which favours the gene for altruism because these workers and reproductives have a 50% chance of inheriting a copy of the same gene for altruism from their mother. The importance of kinship to Hamilton's explanation of altruism led to the theory being dubbed 'kin selection'

(Maynard Smith 1964), although kinship was only ever one way that individuals may come to share genes (Hamilton 1964b; see Chapter 4). Nonetheless, the way that individuals may share genes based on their kinship preserves an individual-centred account of evolution by natural selection because a trait can be explained as an adaptation without needing to know which individuals have particular genes—even though traits are favoured by natural selection because of their contributions to their causal genes' replication.

Inclusive fitness theory is much more than an explanation of altruism by kin selection because it identified what traits are designed to do, which is often explained through a fallacy. Around the same time that Hamilton was founding the notion of inclusive fitness, Vero Wynne-Edwards (1962; 1986) and others were developing an alternative idea of group selection, where Darwin's argument for natural selection among individuals was re-interpreted to be overridingly among groups. Wynne-Edwards (1962, p. 20) argued that

> Survival is the supreme prize in evolution; and there is consequently great scope for selection between local groups . . .. Some prove to be better adapted socially and individually than others, and tend to outlive them, and sooner or later to spread and multiply by colonising the ground vacated by less successfully neighbouring communities. Evolution at this level can be ascribed, therefore, to what is here termed group-selection—still an intraspecific process, and, for everything concerning population dynamics, much more important than selection at the individual level.

Making use of Hamilton's logic, George Williams (1966) argued against such group selection in *Adaptation and Natural Selection*. Williams (1966, e.g. pp. 122–124) concedes that group selection may certainly occur, but that it is always weak compared to individual selection because of the constraints that the genetic mechanism of inheritance imposes on evolution. Quite simply, a mutation could arise that could cause an individual to exploit group-beneficial traits, which would spread throughout the population because it could gain all the benefits of the group-beneficial traits from those in its group whilst not paying for its costs by displaying the group-beneficial trait themselves. In this way,

Williams advocated abandoning the logic of group selection because it can be misleading and, instead, endorsed an individual-based argument because, unavoidably, individuals are responsible for the differential passage of heritable variation through reproduction. Williams (1966, pp. 200–203) was a little sceptical about the experimental utility of inclusive fitness given the limited evidence at the time (see also Wilson 1975b, p. 120), but nonetheless had no trouble with the logic because, critically, inclusive fitness was an individual-based method of accounting the selection for a trait on a gene, and so would only suggest that a trait would be favoured by selection if that trait could evolve within the system of genetic inheritance.

To consider how genetics constrains trait evolution, imagine a group-defence behaviour that leads to an individual sacrificing their own life to keep their group safe, which is seen during colony defence in honeybee stings or exploding ants. For these behaviours to be favoured, there has to be a way for the gene that causes an individual to sacrifice their life to benefit from their sacrificial group defence. Otherwise, no matter how much of a competitive advantage it provides to a group, it cannot be favoured within the system of genetic inheritance. As such, there may be many wonderful adaptations that would help individuals to survive and reproduce, but only those that help the genes that cause them to replicate into the next generations can be favoured by natural selection. Accordingly, benefitting the gene requires the other members of the group to be sufficiently more likely to share an altruistic allele with the sacrificed individual than have a competing allele, which is the case if groups are largely made up of close kin—as, indeed, is the case for honey bees and exploding ants. This simple point caused a considerable amount of confusion because of the enthusiastic pronouncement that group selection always provides faulty logic, whereas hindsight has revealed the more nuanced claim that the only kind of group selection that works (as was also being correctly formulated at the time of Vero Wynne-Edwards's writing by David Wilson (1975a), Michael Wade (1978), and others) is the kind that is consistent with the inclusive fitness explanation of kin selection in involving genetically related individuals.

Even among more self-consistent forms of group selection, inclusive fitness is uniquely an explanation of what traits are designed to do (West and Gardner 2013). There are many ways that natural selection can be accounted by evolutionary biologists to accurately predict evolutionary change. The simplest form of book-keeping is simply to record the number of offspring produced by each individual and their genetic makeup. Inclusive fitness is a more complicated way of book-keeping because it requires the calculation of the marginal increase in the number of offspring produced as a result of a social behaviour, which can easily lead to mistakenly double-counting the benefit to others (without stripping out the same benefit that is received from others by the actor in the calculation of the cost to the actor). But, when correctly formulated, inclusive fitness makes the same prediction as the individual accounting method—and so too does a correctly formulated group selection, which partitions fitness effects into those arising within and among groups. But neither an individual nor a group is in control of their fitness because part of it arises from the social environment of the individual, which depends on what other individuals do. On the other hand, inclusive fitness is the consequence of the traits that an individual has because of its genes— and therefore describes the effect of an individual expressing a trait because of a gene that it has on the replication of that gene (and its allelic copies). Therefore, inclusive fitness is a special kind of causal accounting of genetic change by describing what a gene has control over, and so its correspondence between individual traits and gene replication ensures that it never makes an incorrect prediction about evolutionary change.

Whilst inclusive fitness is individual-centred like Darwin's original account of natural selection, it shows a keen awareness that the traits produced by natural selection are only explicable with an understanding of genes. As such, Hamilton broke with the narrative thrust of *The Modern Synthesis*, but this was only obvious in Hamilton's reasoning—because the result of inclusive fitness justified a reformed account of natural selection on individuals. Before his full statement of inclusive fitness, Hamilton (1963, pp. 353–355) very clearly shows a radical departure in reasoning in a few brief statements:

As a simple but admittedly crude model we may imagine a pair of genes g and G such that G tends to cause some kind of altruistic behavior while g is null. Despite the principle of 'survival of the fittest' the ultimate criterion which determines whether G will spread is not whether the behavior is to the benefit of the behaver but whether it is to the benefit of the gene G; and this will be the case if the average net result of the behavior is to add to the gene-pool a handful of genes containing G in higher concentration than does the gene-pool itself.

This first description of Hamilton's explanation of altruism takes a gene's-eye view of evolution.

The relationship between the gene's-eye view of evolution and inclusive fitness has been a source of confusion, but the two perspectives are obviously different with appreciation of how perspectives have changed over time. Dawkins (1982, p. 194) was very clear about this in *The Extended Phenotype*, which remains one of the clearest expositions of the gene's-eye view of evolution:

> Before Hamilton's revolution, our world was peopled by individual organisms working single-mindedly to keep themselves alive and to have children. In those days it was natural to measure success in this undertaking at the level of the individual organism. Hamilton changed all that but unfortunately, instead of following his ideas through to their logical conclusion and sweeping the individual organism from its pedestal as notional agent of maximization, he exerted his genius in devising a means of rescuing the individual. He could have persisted in saying: gene survival is what matters; let us examine what a gene would have to do in order to propagate copies of itself. Instead he, in effect, said: gene survival is what matters; what is the minimum change we have to make to our old view of what individuals must do, in order that we may cling on to our idea of the individual as the unit of action? The result—inclusive fitness—was technically correct, but complicated and easy to misunderstand.

Dawkins's perspective was heavily influenced by Williams (1966, p. 251), which provided the first systematic account of what would now be recognised as the gene's-eye view of evolution, where evolution by natural selection is described in a way that places the gene at the centre of evolutionary change: 'A gene is selected on one basis only, its average effectiveness in producing individuals able to maximise the gene's representation in future generations. The actual events in this process are endlessly complex, and the resulting adaptations exceedingly diverse, but the essential features are everywhere the same'. The logic expressed by Williams (1966) was persuasively and originally presented by Dawkins (1976, pp. 15–19) in *The Selfish Gene*, to clarify that the units of inheritance are the true beneficiaries of natural selection, even though natural selection acts by favouring individuals based on their traits. Dawkins extolled the essential insight from Hamilton (1963) that natural selection can favour altruism between individuals, where an actor pays a cost to provide a benefit to a recipient, because individuals are simply vehicles for transporting genes between the generations. From a gene's perspective, if it helps a copy of itself to replicate then it is all the same to its selection as if it replicated itself. In this way, Dawkins pushed Williams's reasoning to its logical conclusion (see Dawkins 1976, p. 11).

Dawkins (1976, 1982) advocated the gene's-eye view of evolution, where the world is stripped of all the aspects that have only a passing importance to focus only what causes evolutionary change in the success of different genes. Under natural selection, genes are playing a replication game, which favours genes that produce traits in individuals that maximise the number of copies of the causal gene that individuals with that gene leave behind in their descendants. Accordingly, the gene's-eye view of evolution begs the question: if I were an omniscient gene, what strategy would I use to win in the replication game? The answer, in short, is that 'a predominant quality to be expected in a successful gene is ruthless selfishness' (Dawkins 1976, p. 2) in pursuit of its own replication. As such, genes are often described as 'selfish' because they are only trying to maximise the number of copies of themselves that they produce. Seeing the world through the lens of selfish genes focuses the attention on what counts for the evolution by natural selection, which is simply how many times a gene is replicated compared to its slightly different allelic competitors, and the rest of what traits do or how evolution proceeds goes into the footnote.

Herein, there is an often unacknowledged alteration in the gene's-eye view of evolution as opposed to the individualistic account from Darwin and Hamilton, encapsulated in how 'Natural selection would produce or maintain adaptation as a matter of definition' (Williams 1966, p. 25). Darwin (1859) gave an account to explain the origin of adaptations through evolution by natural selection, which was a contrasting perspective to the near-mainstream view presented by Paley (1802) of divine Creation. In this way, adaptations were something surprising that needed to be explained, whereas, to Williams and Dawkins, the question was different because adaptations were simply the result of natural selection. Instead, much like the direct and indirect fitness partition of Hamilton's inclusive fitness, the problem to be explained was how a particular adaptation benefits the genes that cause it. The route to benefit the gene was particularly important when, in examples of apparently disadvantageous traits like altruistic behaviour that were the popular focus of attention in behavioural ecology, this was not immediately obvious (as it was for classic traits that promote individual survival). The reframed question underlies the current mainstream view of the design principle of natural selection leading individuals to bear traits that are adaptations as if 'to maximize inclusive fitness' (West and Gardner 2013).

So, now, what does evolutionary genetics through the gene's-eye view add to the understanding of the difference between Kettlewell's (1958) and Allison's (1954) studies? Narratively, the contribution could be cast as either inconsequential or enormous. On one hand, the gene's-eye view adds very little that was not obvious from transmission genetics alone; namely, that the traits are likely to be underpinned by genes that need to benefit from the trait they produce to be favoured by natural selection. But, by emphasising this point, the gene's-eye view of evolution expands the possible explanations of providing an advantage to an individual, and limits it in other ways. For example, similar to Fisher (1930, pp. 160–162), if dark-coloured peppered moths were poisonous, a new colouration could help predators to learn that dark-coloured peppered moths should not be eaten; but such a group selection argument may be unlikely to work because a dark-coloured peppered moth could arise that does not pay the cost of producing the poison. For dark colouration to be a 'warning colour', poisonous moths would need to be more likely to be associated with their kin, such that the protective effect of a poisonous moth being eaten and a predator learning not to do that again mostly benefits individuals that share the gene for the poisonous dark colouration. It should be clear: this is a hypothetical example to show how the gene's-eye view could add something to the understanding of the trait, though it does fit with the decline in dark-coloured peppered moths after urban environments became less sooty (and so camouflage seems to be the better-supported explanation). Yet, for all that the gene's-eye view could add something, if melanism in peppered moths is contrasted against protection from malaria in humans, little has been gained towards understanding why natural selection did what was expected in in one case and not in the other. Accordingly, as well as understanding the logical structure of the inheritance of traits in transmission genetics and how traits are favoured by natural selection in evolutionary genetics, it is also necessary to understand what genes physically are, how they produce traits, and, altogether, what really changes when there is evolution by natural selection.

* * *

Molecular genetics describes how traits are produced from heritable material. The foundational experiments that demonstrated the properties of the molecule of inheritance and that DNA is that molecule are fascinating, but for all their ingenuity they are also intricate, so it is left for summaries like Horace Judson's (1979) *The Eighth Day of Creation*. Cutting through swathes of biochemistry to a signpost of the start of molecular genetics, the race to understand how DNA produces traits began with the famous discovery of the structure of DNA by James Watson and Francis Crick (1953)—and Rosalind Franklin. DNA is a complicated polymer made up of a small set of nucleotide units, which are referred to by their abbreviated chemical names as A, C, G, and T. The nucleotides are often understood as providing a four-base syntax that stores coded information as a sequence, much like how a computer stores information in the binary code of 1s and 0s. But in moving from syntax to semantics, a large number of people and experiments were then involved in establishing the meaning of the information in the relationship between DNA and traits, which speaks to the professionalisation of modern science. At first, it was initially assumed that traits were primarily the product of proteins, which are large molecules that are themselves polymers made up of a sequence of up to 20 amino acids, where three consecutive bases of DNA specify one element in a sequence of amino acids that folds up into a protein. This specification occurs indirectly, where cellular machinery within the nucleus of a cell transcribes DNA into messenger RNA (mRNA) that leaves the nucleus to be translated into an amino acid sequence that forms a protein at the ribosome. As such, genes were assumed to be synonymous with protein-coding DNA because proteins form the cellular machinery that leads to traits, but this description is totally inadequate.

The definition of a gene as a protein-coding region of DNA can cause a little confusion, but molecular genetics has largely moved on from this definition because the assumption of 'one gene—one protein' is incorrect. Non-protein-coding DNA may influence traits through a vast number of molecular mechanisms, including (but not limited to): influencing the accessibility of cellular machinery to DNA in the nucleus by affecting the three-dimensional structure of DNA, encoding RNA markers that are targeted by cellular machinery that regulates the decomposition of mRNA within the cell, and encoding other signals that are targeted by cellular machinery that regulates the chemical modification and placement of proteins after their translation by the ribosome. Therefore, beyond the initial suggestion of a gene as a protein-coding region of DNA, contemporary molecular biologists tend to refer to a gene as any stretch of DNA that serves as a discrete unit of function in bringing about its impact on traits. There is no need to be pedantic about exactly how long a stretch of DNA need be to 'count' as a gene, as it would depend on the focus at hand: when discussing the evolution of transcriptional control it is often useful to talk about genes as quite small stretches of just a few bases of regulatory (non-coding) DNA, whereas when discussing the evolution of multicellular development it is often more useful to talk about genes as thousands of bases of DNA spanning both protein-coding and regulatory stretches. Herein, a gene is a label used as a tool in the service of research.

In the previous section, an alternative concept of a gene was used from evolutionary biology, which was historically traced from its original usage to describe Mendel's hypothetical particle of inheritance through to its more recent and famous usage by Dawkins (1976). Intentionally, how Dawkins (1976, p. 28) defines a gene was not directly addressed, which lifts a definition from Williams (1966, p. 24) as 'that which segregates and recombines with appreciable frequency'. The central idea here is that sexual recombination—from both the independent assortment of chromosomes and the crossing-over of genetic material on homologous chromosomes—breaks apart gene combinations between the generations during meiosis, and so the gene is whatever unit of inheritance is constant throughout the process of mixing. Like with the definition from molecular genetics, the purpose of this definition is to provide flexibility to handle different problems that could be addressed; hence Dawkins suggests that a gene could be of almost any length. Consequently, with the discussion of altruism in *The Selfish Gene*, Dawkins (1976, p. 32) describes a gene as a stretch of DNA that is 'small enough to last for a large number of generations and to be distributed around in the form of copies'. As Williams (1966, p. 24) explicitly linked, this meaning of gene is consistent with the concept from Mendel that is used in population genetics. With this definition it is worthy of note that whilst molecular genetics has revealed that hereditary material is physically substantiated by DNA, the gene's-eye view of evolution has essentially not been updated since 1966 because the population geneticists were so fantastically successful at deducing the logical structure of inheritance (at least, as it is important for evolution) without need for understanding the physical process of inheritance.

The definition of a gene in molecular genetics has been set in contrast with the definition of a gene in evolutionary biology because there is an important distinction between a gene as a unit of inheritance in evolutionary biology and as a unit of function in molecular genetics (Haig 2012). The design of traits is built by the process of evolution, but traits themselves are produced anew in every generation. The way in which a piece of DNA causes a particular trait can be very complicated, through the interaction between the DNA and its cellular, bodily, and wider environment. In this way, it is important to recognise that when geneticists discuss a gene for melanism that gives peppered moths a dark colouration, they are making a statement about the functional difference between alleles and not necessarily about what cellular or bodily functions a gene contributes towards. This may seem like a peculiar way of framing things, but it is how natural selection works: an allele is favoured over its counterpart(s) on the basis of what it does differently and not what is does the same. In the case of melanism in peppered moths, this example is a canonical study of natural selection in wild populations and yet the function of the gene with the mutation that causes melanism remains to be discovered (Cook and Saccheri 2013). In keeping with such logic, David Haig (2012) defends this perspective to make the case that evolutionary biology is justified in only concerning itself with genes as units of inheritance that cause slight differences between traits.

With a view to the discussion of maladaptation in later chapters, the defence is difficult to maintain. But, even more generally, it is dissatisfying for being narrow-minded. The fundamental problem at the heart of evolutionary biology is the appearance of design, which is about how bodies work. Although the explanation of design may involve the gradual modification of traits over the course of evolution by natural selection on alleles based on the slight difference in traits that they produce, this does not get away from the reality of molecular genetics that DNA is packaged into relatively discrete functional units. Of course, functional units can be grouped together through their interactions into systems that could be thought of as functional units in their own right, but this also applies to heritable units. Whatever conceptual scheme is adopted to describe the mechanisms of inheritance and function, at base, heritable units are selected based on their functional consequences. So, even if inheritance and function both have messy realities, some correspondence is expected between genes as units of inheritance and as units of function.

Against this stance, Dawkins (1976, pp. 38–39 and pp. 84–86) famously used a rowing boat analogy to justify the focus on units of inheritance alone in the gene's-eye view of evolution; the idea being that natural selection with free recombination is like a good coach in swapping in and out the best rowers in each seating position to get the fastest boat to win the race. The analogy is consistent with the mathematics of population genetics, where natural selection acts based on the average 'additive' improvement that each allele affords to individuals' fitness (Fisher 1930, pp. 34–35). But the fault can be seen in the rowing boat analogy: a good coach may swap in and out individual rowers, but they would not cut-and-glue rowers together if one rower's left side was best matched with another rower's right side, because both rowers would die—or, in the steely eyes of their coach, become 'less effective' rowers. As such, the argument implicitly assumes that units of inheritance and function are equivalent.

The argument in the spirit of Dawkins (1976, pp. 84–86) might protest that, regardless of rowing boats, this is exactly what natural selection would do with DNA if there were different mutations within an allele that each afford independent improvements, which would seem reasonable given the mathematical support of the analogy. But, when considering the diversity of possibilities for real genes, the salient point is that it does not have to be the case: natural selection can favour an allele because of its combined effects with other mutations (Wright 1931), which has been known to fuel the diversification of populations in different localities. An obvious case of this arises in the evolution of pesticide resistance, where once a resistance mutation has spread through a local population, a resistance mutation with the same effect from another population is unlikely to spread. For example, *Anopheles* mosquitoes have evolved target-site resistance to pyrethroids that are used to control their numbers (and limit the spread of the malaria that they carry), but mutational variants have arisen at different localities, which has led to particular mutations being associated with eastern and western parts of sub-Saharan Africa (Clarkson et al. 2021).

Moving on from rowing boats, the argument for treating a gene as a unit of inheritance and function can be further considered on its own terms. Imagine two alleles that are separated by two mutations in a protein-coding region that both influence a protein's functionality, such that these two alleles are selected for producing different traits. If recombination within the protein-coding region were to break apart the two mutations that separate these two alleles, what would happen? The mismatched mutations could produce a protein that affords a trait unlike either of the existing traits, and so would intuitively be treated as a new allele because the combination of mutations leads to a different function. Alternatively, the mismatched mutations could produce a hybrid trait that acts like a functional intermediate. So all the options are on the table. The question then becomes: how would natural selection act?

In general, it is reasonable to assume that recombination within the protein-coding region would be rare compared to the effect of selection on the frequency of the combination of mutations. After all, natural selection occurs every generation, whereas recombination would be extremely unlikely to occur in exactly the same place every generation to break apart the protein-coding region. Consequently, the fate of this new combination of mutations critically depends on its change in frequency from selection, as it is not a free agent under the expected frequency of recombination. Further, the fact that the two mutations contribute to the same protein does necessarily mean they will be selected together; there is no way for the two mutations to produce independent functions that could be selected independently, for instance due to independent expression, because the protein is a discrete entity with a functionality that necessarily arises from the action of the two mutations together. As such, it makes intuitive sense to treat the new combination of mutations as forming a new allele, rather than treating each of the two mutations as separate alleles at different loci.

There is potentially much to debate about the general utility of this perspective. But, for the purposes here, it is critical to see this perspective as a 'thinking tool' that is developed for a purpose. Just as Dawkins (1976, p. 32) identified the gene as a small unit of inheritance for the purpose of understanding Hamilton's (1963) logic of altruistic behaviour (where genes need to be identically shared between individuals), which is not disputed, so too here the gene is identified as a unit of inheritance and function for the purpose of understanding Allison's (1954) logic behind the explanation of the intermediate-frequency protection against malaria (and, more broadly in later chapters, maladaptation). It makes sense to recognise that DNA really does come packaged in relatively discrete units of function, even if those units of function can vary in size (up to e.g. supergenes or sex chromosomes), because it is the reality of how DNA is selected and how it produces traits; it is central to the problematic difference between Kettlewell's (1958) study of industrial melanism in peppered moths and Allison's (1954) study of protection against malaria and sickle-cell anaemia. The purpose of acknowledging this reality is to open wide a conceptual distinction that explains why protection against malaria might not spread throughout the human population in the tropics on account of its association with sickle-cell anaemia.

Some traits are extrinsically pleiotropic, meaning that the association between traits is coincidentally related to factors other than the traits themselves. There are a variety of possible reasons behind extrinsic pleiotropy that in one way or another arise from historical contingency. Perhaps the most obvious is association from physical linkage due to proximity on a chromosome. If this were the case in the association between protection against malaria and sickle-cell anaemia, their association could simply be because alleles for these two traits are co-inherited because they are found very close together. Consequently, the association between these traits would be because recombination is unlikely to break apart the allele combination. The selection of these traits together might reflect two competing selective pressures: selection for malaria protection in parents and selection against sickle-cell anaemia in offspring. The spread of a deleterious allele through physical linkage to a beneficial allele is sometimes referred to as hitchhiking (Maynard Smith and Haigh 1974). Given that protection against malaria is dominant and the sickle-cell anaemia is recessive, the allele combination would reach intermediate frequency as the chance of sickle-cell anaemia affecting offspring only becomes appreciable when both parents carry the sickle-cell anaemia allele. Whilst an intermediate frequency of the allele combination might persist over many human generations, recombination would be expected to break apart these alleles over longer timescales. In this way, extrinsic pleiotropy may slow the evolutionary process of design improvement from natural selection, but it is not expected to change the end result of which traits spread.

Other traits may be associated together by intrinsic pleiotropy, which would mean that the association between protection against malaria and sickle-cell anaemia would be a more permanent state of affairs. If a single mutation were responsible for both protection against malaria and sickle-cell anaemia, these two traits are inextricably associated from the common functional role of the gene. Pleiotropy describes the situation where a single gene has multiple functional effects. Whilst it is true that factors like physical linkage may often be cited as a cause of pleiotropy for explaining some trait combinations, there is a fundamental difference between extrinsic and intrinsic pleiotropy in the long run: if the two traits are intrinsically pleiotropic in arising from the same gene function, then the evolutionary fate of those traits is always going to be one and the same. This is not to say that other genes could also influence intrinsically pleiotropic traits, but rather to be clear about what natural selection actually favours when it produces an evolutionary change.

As Cronin (1991, p. 107) incisively recognised, pleiotropy is a major reason why natural selection is fundamentally gene-centred, whether or not a gene's-eye view of evolution is taken:

> If a genetic change that lengthens the bone also curves the eyebrow, then our adaptive explanation should recognise that; we should be interested in the genetic differences that give rise not merely to differences in toe-length but to differences in toe-length-plus-eyebrow-shape, even if eyebrow shape should turn out to be selectively neutral. This is an answer that would not have been obvious to the organism-centred view of classical Darwinism but comes readily to a theory that is gene-centred.

The critical point is that without a gene-centred understanding of natural selection, pleiotropy appears to occur arbitrarily. With a functional understanding of genes from molecular genetics, pleiotropy can become a logical consequence of how a gene causes its traits. Moreover, a functional understanding of traits enables the separation extrinsic and intrinsic pleiotropy, which helps us to see what natural selection is really promoting.

With sympathy for evolutionary geneticists, the work has not always been done to have a clear picture of the molecular details of how genes cause traits. Nonetheless, a basic understanding can be revolutionary for understanding the evolution of a trait. Cronin (1991, p. 99) offers an important demonstration of this in thinking about how a small genetic change could produce an amazingly complex suite of adaptive changes in organismal traits. A gene that increases the head size of a bison must also need to increase the size of the neck muscles to allow the bison to manoeuvre its head. Whilst this might require 'a quirky stroke of luck' for the gene to bring about both changes, through 'extended pleiotropy' a 'large-headed bison will automatically tend to develop large neck muscles' because a bison has 'a tendency for those muscles to grow if they are exercised' (Cronin 1991, p. 99). So, instead of appearing insuperable that natural selection for a larger head would also need to be very intimately associated with a gene for larger muscles, the problem vanishes when it is realised that there is a logical connection between the traits. In this case, a single genetic change in head size may be sufficient to bring about all the pleiotropic changes needed for its natural selection.

Such logic can be exercised in the case of sickle-cell anaemia and protection against malaria to understand how intrinsic pleiotropy is the cause of its differing evolutionary trajectory. Functional analysis using the techniques of molecular genetics has established that the allele for the conjoined trait arises from a single nucleotide mutation in the protein-coding sequence for beta-haemoglobin, which is a protein involved in oxygen transportation around the body in the blood (Ingram 2004). The mutation replaces the sequence GAG with GTG for the sixth amino acid position, leading to valine replacing glutamic acid in the amino acid sequence. This alteration causes the beta-haemoglobin protein to fold up in a different conformation that can distort the shape of a red blood cell (into a characteristic sickle shape), which makes the cell more unstable and so more likely to rupture as it is pumped around the circulatory system. As such, the A to T mutation creates a loss of function in making red blood cells less capable of oxygen transport. But there is also a curious gain of function because the instability also means that, when a malaria parasite tries to invade a red blood cell to find a safe location within the body to reproduce, the red blood cell is more likely to rupture. This releases the malaria parasite back into the blood plasma, where the white blood cells of the immune system can attack it. Accordingly, the mutation that causes sickle-cell anaemia also confers protection against malaria because of its same functional consequences in destabilising red blood cells. Consequently, it seems unlikely that additional genetic change could cause malaria protection in parents without the risk of sickle-cell anaemia in their children.

Therefore, with insights from transmission genetics, evolutionary genetics, and molecular genetics, the fundamental difference between Kettlewell's (1958) study of industrial melanism in peppered moths and Allison's (1954) study of protection against malaria in humans can be pinned to intrinsic pleiotropy; a great deal of knowledge to reach a relatively simple conclusion. But the task of this book is not to explain adaptation or non-adaptation in either of these cases; instead, the problem is to understand maladaptation. Analogously to protection against malaria and sickle-cell anaemia, the reasoning about intrinsic pleiotropy is key to beginning to understand maladaptation. A maladaptation is a trait that is harmful to individual survival and reproduction. If that were all it was, it would be difficult to see how it could be favoured by natural selection. But if there is a logical relationship to other consequences, then, as the next chapter explores, the way towards understanding maladaptation begins to become clearer.

3

Population ecology of natural selection

The case from the previous chapter that it is necessary to incorporate genetics into evolutionary theory to understand how natural selection really works is already well-established in the minds of evolutionary biologists. There is also a need to incorporate ecology into evolutionary theory. This remains somewhat contentious, so it is necessary to be clear about the minimal claim that is being made. Just as the majority of the important insights from genetics for evolutionary theory would naturally fall under the subheading of 'population genetics', so too here ecology is meant primarily in the sense of 'population ecology'—principally in reference to understanding how evolution by natural selection impacts a species' population size. This is a very limited sense of ecology, which is all that is needed here. More generally, there is a need for much more to be done to incorporate ecology into evolutionary theory than merely exploring the effect of natural selection on population size; Andrew Hendry (2016) has already made an admirable case for this in *Eco-evolutionary Dynamics*, which will not be repeated. Nonetheless, to understand maladaptation, evolutionary biology must appeal to population ecology. The argument falls into three parts: first, establishing the logical possibility of maladaptation (that is more rigorously established in a mathematical appendix); second, clarifying why maladaptation has been viewed as impossible; and third, reflecting on why the concept of maladaptation is needed.

Starting with its logical possibility, maladaptation arises from a fundamental disharmony between the ecology and genetics of evolution by natural selection. Under Darwin's pre-genetic understanding, natural selection favours those individuals that are better at surviving and reproducing, which (as is elaborated on later) has an obvious ecological expectation of increasing population. The details of the genetic mechanism of inheritance transform the understanding of how, why, and what is actually selected by natural selection. A gene is favoured by natural selection for its ability to increase its frequency. It may be hard to spot the difference here, but there is an important distinction: the maximand—the thing that natural selection acts to increase—is changed from an absolute to a relative measure of success. Population has no upper limit, whilst frequency is proportional, and so is constrained to have a maximum of one (meaning that every individual has the same allele at its locus regardless of how many individuals there are). Most evolutionary biologists take no heed, and continue to say thing like 'individuals are selected to maximize the number of surviving offspring they produce' (Trivers 1985, p. 21). But the distinction between absolute and relative success opens up the possibility of maladaptation, where a gene contributes towards a trait in an individual organism that decreases its absolute copy number among individuals, but increases its relative rate of replication into them. Strategically, a gene may only benefit from harming its absolute success if, by doing so, it increases its relative success by delivering more harm to competing genes. How can this occur?

The explanation can be established without appeal to extraneous biological details, not least because increasing relative success by compromising absolute success presents a familiarly human tragedy across a variety of activities. Perhaps this is most clearly experienced in the amoral and closed-system context of competitive games. Consider two Olympic sports (in keeping with the logic made explicit in Ghiselin 1974, pp. 49–51; Sober 1984, pp. 15–17). In archery, two players shoot arrows at a target in sets of three, where the archer that shoots closer to the bullseye than their opponent wins the set, and the archer that wins the most sets wins the match. Archers cannot interfere with each other's shooting, and so the only way that a player can increase their chances of winning is by shooting more accurately. Although an archer's progression through the tournament depends on scoring better than their opponent in each match, the competition setup favours archers based on an absolute standard of success in their ability to shoot closer to the bullseye. By contrast, in boxing, two players enter the ring and fight each other for a series of rounds, where boxers are awarded points for landing clean blows on their opponents, and the boxer with the most points at the end wins. In this competition setup, players cannot be meaningfully assessed on an absolute standard of success because it is all about doing *relatively* better than an opponent. Indeed, a point to one boxer could equally be scored as a negative point to the other. It does not matter if a boxer scores two points or twenty to boost their own performance; boxers need only focus on landing more clean blows than they take.

The two kinds of success in the sports can be thought of in units of fitness in nature. Much has been written on the nature of fitness, but very little has been said. Following Fisher (1930, p. 34), fitness is a propensity that can apply to whole individuals or the marginal effects of alleles. Regardless of the subtleties of its definition or measurement, either form of fitness can be framed as an absolute or relative metric. Absolute fitness can be broadly conceived in the units of population, such as the number of surviving offspring. Relative fitness can be likewise conceived in the units of frequency, often expressed as a rate of logistic increase.

Natural selection relies upon an allele having higher relative fitness than a competing allele at its locus, which may lead to adaptation or maladaptation depending on its effects on absolute fitness. Consequently, like a boxer at the Olympics, an allele is favoured by natural selection through its relative performance. But the allele also competes like an archer when this corresponds to increasing absolute fitness (e.g. Fisher 1930, p. 42). The direction of selection for an allele for maladaptation that obtains higher relative fitness through lowering absolute fitness is a test case that can distinguish whether nature is more like a boxing ring or an archery range. This sounds simple enough, but the struggle for existence in nature is unlike competing at the Olympics in that there is no

oversight to enforce the rules. As such, the rules in the game of life—
in as far as there are any—are set by the environment. So: what about
the environment can change the nature of the competition? Again, it is
useful to seek intuition by considering another Olympic sport, but this
time one where the standard of success is more open-ended than boxing
or archery.

In long-distance running, players often have to compete by qualification, a series of heats and then a final. In qualification, a runner needs
to demonstrate that they can run the distance in a time that meets the
competition standard to be entered into the heats. This is often done by
runners completing the track on their own. In the heats, runners need
to complete the track in a top-ranking time to go through to the final.
Once they get to the final, medals are awarded to runners that complete
the track in the fastest time. The competition setup for long-distance
running may sound superficially similar to archery in always favouring
runners that complete the track in the *absolutely* fastest times, but there
is a key difference between qualification, the heats, and the final.

Imagine a savvy runner that is considering a new strategy for long-
distance running that counter-intuitively involves a degree of specialising
in sprinting. By focusing some of their training on sprinting, they would
be able to run faster than anyone else over a short distance, but, by
neglecting their endurance training, they would be slower than others
over a longer distance. Intuitively, it might be supposed that this sprint-
specialism would be unlikely to win the runner the gold medal, but they
have a plan. By sprinting to the front of the pack at the start of the race,
they can get ahead of the competition and have some control over the
speed of the pack because it becomes more difficult for others to over-
take them later in the race—especially on the corners. The savvy runner
may not be able to slow the pack down indefinitely, but as long as they
can slow the endurance-specialists down enough, the savvy runner will
have enough energy left to sprint past them in the final stages of the
race. Indeed, the savvy runner may elect to do this by speeding up on
the straights of the track where it is easier to be overtaken and only slowing down on the corners where it is harder to be overtaken. Through

this plan, the savvy runner can complete the track relatively faster than the competitors in their race, but they are unlikely to set a new world record. The plan should work well in the final because the medals are awarded based on ranked times, but it would be of no help for qualification because there is no one else there to slow down. In the heats, it is more complicated assessing whether or not this would work: by slowing down themselves and their competitors in their race, the savvy runner may well come first in their heat, but qualification for the final depends on their time relative to competitors in other races as well. The success of their strategy would depend on who the savvy runner is competing against more: those in their race or those in other races. Therefore, overall, the competition setup of the final could be likened to a boxing ring, qualification to an archery range, and the heats to somewhere in between.

Such ambiguous competition in sports is closer to how competition is in nature. It is easy to think that an individual organism is competing against all members of a population, but the individual exists at a locality, which limits who their competitors are. Consequently, regardless of theoretical assumptions, it is rarely true in reality that individuals that are far away from one another are in competition for resources, mates, or anything else. This has received the most attention in evolutionary biology in the study of sex allocation (West 2009), where the key parameter that varies the incentives in a race between the extremes of an archery range and a boxing ring has been referred to as 'the scale of competition' (Frank 1998, pp. 114–115). In the final or with a small number of parallel heats, the savvy runner is only competing at a *local* scale against those other individuals in their race, so the pack-slowing strategy is likely to succeed. Alternatively, in qualification or with a larger number of parallel heats, the savvy runner is competing at a *global* scale against a larger number of individuals in other races, so the pack-slowing strategy would be likely to fail. In nature, localised competition most obviously occurs among individuals when it arises at the scale of the group. An individual is likely to have some control of the survival and reproduction of their groupmates because they can interact with them due to their proximity, whilst they have almost no control over individuals in other groups in the same population that may be very far away. Just like for the savvy runner slowing the pack in their race, when competition occurs at the local scale of the group, an individual can benefit from a maladaptation through performing relatively better than other group members by harming their own success if it also causes sufficient harm to their competitors' fitness.

In nature, competition is different to how it is in sports. To cause sufficient harm, there is a key problem for a gene to overcome that no athlete ever encounters. When a new allele for a maladaptation first arises by mutation from a wild-type allele, it is very rare within a population. For the mutant allele to have a meaningful chance of spreading to higher frequencies, it must have higher fitness relative to other alleles to be favoured by natural selection. For simplicity, the allele for maladaptation can be considered as competing against a single wild-type allele, which is identical except that it does not produce the maladaptation. The relative fitness of the wild-type allele is frequency dependent because the frequency at which maladaptive harm to individual fitness occurs is tied to the frequency of the mutant allele. This can make things complicated because natural selection is also frequency dependent.

Under global competition against a very large number of individuals, a mutant allele for maladaptation is unlikely to spread because of its frequency-dependent relative fitness (Madgwick 2020). The mutant allele starts out as very rare, meaning that it only very rarely harms the wild-type allele, and so the wild-type allele's fitness is almost unchanged by the mutant allele. By contrast, the mutant allele harms itself all the time. As a result, the mutant allele is unlikely to have higher relative fitness. The intuition for this has already been outlined in the case of the heats, where the savvy runner is not obviously able to benefit from the pack-slowing strategy because all the runners in the savvy runner's race are likely to be outcompeted by runners from other races. If global competition occurs in nature, it is as if there are a very large number of races running in parallel, such that the competition generally favours individuals based on their absolute fitness.

It is widely argued by evolutionary biologists from diverse traditions that global competition is not the norm in nature because populations are spatially structured (as reviewed in Wright 1969). The assumption of global competition is nonetheless widely made in evolutionary analyses—not because it is true, but because of a bias against incorporating the complexity of population structure into mathematical models. Regardless of what plausible spatial model is used, population structure lends itself to localised competition because individuals that are far away are not competing with the intensity of individuals that are close together. Moreover, these individuals tend to be genetically similar due to limited dispersal; in other words, offspring tend to live closer to their parents, rather than randomly disperse within the range of the population's habitat, not least because the offspring were produced at their parents' location and movement can be costly. So the salient biological question is not whether populations have localised competition, but how small the group is where localised competition occurs.

Under local competition against a very small number of individuals, a mutant allele for maladaptation is more likely to spread. Localised competition has what is going to be called an 'evolutionary effect', whereby a single mutant allele would have a higher frequency by a straightforward inflation from being counted among fewer individuals (Gardner and West 2004, p. 1197): one allele among ten has a higher frequency than one allele among thousands. Such inflation could tip the balance of selection to favour the mutant allele when it is very rare. However, to make a big difference, there would have to be only a few individuals, and it is not obvious that such small groups regularly characterise the scale of competition in nature (Hamilton 1971). On the basis of this effect alone, maladaptation cannot easily evolve, which some proponents of forerunning ideas relating to maladaptation overlook.

Localised competition also has an 'ecological effect' that applies to larger groups of individuals, which is the more complicated result of population feedbacks. Early evolutionary analyses of harmful social traits (e.g. Hamilton 1970; Grafen 1985) assumed that there were two classes of individuals to consider: the actor who harms their own fitness and the recipient whose fitness is harmed by the actor. But there is a third party that also needs to taken into account, which came to light in analyses that explicitly consider population ecology (e.g. Taylor 1992; Wilson et al. 1992). Besides the actor and the recipient, there is also a benefi-ciary that is constituted of the remaining individuals that are not part of the interaction but are nonetheless impacted by it, as absent members of the same local population. These individuals have already featured in the consideration of the savvy runner; in the long-distance running anal-ogy, this would be those individuals in other races that stand to benefit from the pack-slowing strategy. Therefore, although it is an ecological effect, it is a short-term one that impacts absolute fitness within a gener-ation, rather than relying on any long-term feedbacks on population size over many generations, which are very unlikely to impact the direction of natural selection as individuals move away from the long-term effects that they cause (Trivers 1985, p. 83). Long-term feedbacks may lead to similar outcomes *in extremis* like population extinction, but such 'evo-lutionary suicide' (Parvinen 2005) is an entirely separate phenomenon to maladaptation.

Obscuring the ecological effect, another pernicious assumption of mathematical models in evolutionary biology is that populations are 'inelastic' (Taylor 1992), so that evolutionary change can take place with-out impacting population size (at any scale). Again, this assumption is obviously false, but it is convenient. It is perhaps most valid for models with weak selection, when the small change in fitness can be neglected for the sake of simplicity. Yet, such approximations may also neglect the ecological consequences of evolutionary change, which may impact the direction of evolution—as, indeed, it does for maladaptation.

The ecological effect is nothing surprising to population ecologists. Much like the number of places for runners in the final, an ecological niche can be imagined as a resource of a fixed size at a locality. When an allele for maladaptation causes an actor to pay a cost to harm a recipi-ent, both the actor and recipient have reduced absolute fitness in a lower probability of surviving to reproduce. If one or other of those individuals dies (or is less capable of obtaining that resource), more of the resource is freed for the use of others at the locality, which increases their prob-ability of surviving to reproduce. The effect is nothing controversial, as it is a basic consequence of density-dependent feedbacks (Frank 1998, pp. 114–115) or something equivalent (e.g. reproductive compensation). It is an ecological (rather than evolutionary) effect in the sense that the feedback is not conditional on the genetic identity of the individuals that benefit from it, so it does not inherently favour an individual with one allele or other. Instead, the ecological effect is an immediate consequence of more resources becoming available for individuals.

Whilst there is no inherent bias to the ecological effect, it can have evolutionary consequences when other factors are considered. Through limited dispersal, population structure means that individuals in the same locality are more genetically similar than random members of the whole population. Consequently, the allele for maladaptation may reap some benefit through the reduced competition for resources at the locality. The total benefit to the allele is frequency dependent; if the allele has a higher frequency, then it is more likely to benefit from reduced competition. But the benefit that is returned because of population structure is frequency independent through arising from fixed features of the environment. For example, if individuals are cousins, then they always have a 25% chance of sharing an allele by common ancestry, irrespective of the frequency of that allele in the whole population.

The astute may recognise that population structure also means that the recipient may be more likely than random to share the allele for maladaptation through common ancestry. Indeed, for a mutant allele for maladaptation to be favoured by natural selection, an ancillary requirement must be met. After there being an ecological effect that returns some benefit back through the third party, the recipient must also be less likely than a random member of the local population to carry the allele for maladaptation. This ensures that there is 'negative relatedness' between the actor and the recipient so that the allele for maladaptation makes a relative fitness gain in the local population from the harm that it may (at least sometimes) deliver to itself. Consequently, maladaptation cannot evolve through the indiscriminate harm of random members of the local population (Patel et al. 2020). Yet, there is no need for a sophisticated mechanism to recognise negatively related recipients (as is often assumed e.g. by Patel et al. 2020). Indeed, population structure is ever-present (Wright 1969, p. 290), and an individual need only move away from its location to change its relatedness to those that are nearby in the same local population. For example, a chick may leave its nest to harm its own and its cousins' absolute fitness in a neighbouring nest, which may increase the relative fitness of the causal allele for maladaptation by reducing the competition among its siblings at its own nest.

With the selective advantage of maladaptation depending on fixed features of the population structure, there is no reason to think that it would not spread through the whole population. Yet, so it has been found, there is often reservation in thinking that maladaptation would be able to lead to the extreme possibility of extinction for the whole population. Arguably, whilst it may arise in a local population, either it would drive the local population extinct before it spreads to another locality, or natural selection acting at the level of local population would disfavour maladaptation. Local populations with the allele for maladaptation would be expected to produce fewer migrants than those without, but, due to frequency-independent selection, it would only take a single migrant with the allele for maladaptation for it to have a chance of spreading in another local population. The spread of an allele for maladaptation that drives a local population to extinction would therefore depend on the probability that the allele can escape to a neighbouring population before the declining population becomes extinct. It is arguably possible that the decline could be so rapid that it would not occur (as if by the natural selection at the level of the global population). But the idea of selection operating at the level of global population in this way is misleading because the only way that there would be no chance of the allele for maladaptation not spreading between local populations would be if it drove the local population to become extinct within a single generation (i.e. by reducing absolute fitness to zero), whereupon the trait is not a maladaptation because it cannot be said to be favoured by natural selection anyway (because the relative fitness of its causal allele would also be zero).

Therefore, through the ecological effect of localised competition, the scope for widespread maladaptation opens up. Consequently, there should be little doubt that the population ecology of natural selection is paramount to evolutionary analysis, given that this new possibility for evolution arises from a local decrease in population density. The assumed parity of an allele that causes an absolute fitness increase in population and a relative fitness increase in frequency causes an interaction that has been neglected for too long—and here the phenomenon of maladaptation cannot be understood without recognising the population ecology of natural selection. In this way, maladaptation is a paragon for a population biology that incorporates ecology into the genetic theory of evolution because it is the differentiating example that only makes sense with an integrated understanding of the genetic and ecological basis of evolutionary change.

* * *

Having established a logical case for the possibility of maladaptation, it might appear strange that maladaptation could have been viewed as impossible. Whilst there are some theoretical puzzles to solve before maladaptation appears possible, it almost seems as if maladaptation has been wilfully ignored. The explanation for this lies in the depths of the history of evolutionary biology. Obviously, maladaptation does not get a direct mention, so the explanation must be pieced together from the clues that there are in the implications of weakly related discussions.

Although the neglect of maladaptation extends beyond the foundations of evolutionary biology, a good place to start is with Darwin, asking: what does his theory of evolution by natural selection set out to explain? Despite the title of *The Origin of Species*, when Darwin (1859) first presented his theory of evolution by natural selection, he did not offer an account of the origin of life or the formation of new species out of existing ones. Darwin's claims about these topics are only broached towards the end of the work, where Darwin (1859, pp. 481–484) couches his opinion that species evolve out of one another alongside his 'doubts' and 'difficulties', as an extrapolation that extends beyond the evidence and argument of the book that had him 'fully convinced'. Evolutionary biology would have to wait several more decades for the theories of allopatric and sympatric speciation (as reviewed in Mayr 1982, pp. 561–566). Instead, Darwin (1859, p. 60) asked: 'How have all those exquisite adaptations of one part of the organisation to another part, and to the conditions of life, and of one distinct organic being to another being, been perfected?'. Darwin (1859, p. 480) presented evidence that supports natural selection as the mechanism by which existing 'species have changed, and are still slowly changing by the preservation and accumulation of successive slight favourable variations' to suit their environments. Although he played to the fashionable question of speciation, Darwin really set out to explain adaptation.

It is difficult to pin down what precisely Darwin means by adaptation because he never defines the term. Nonetheless, it is likely that what is here called maladaptation displays a sort of design that lies beyond what Darwin meant as adaptation, and so he would have cast maladaptation as impossible on theoretical grounds. Throughout *The Origin of Species*, Darwin (1859, p. 149) repeats words to the effect that 'Natural selection, it should never be forgotten, can act on each part of each being, solely through and for its advantage'. Elsewhere Darwin (1859, p. 211) couches this in reference to Paley's (1785, pp. 40–41) words in *Moral Philosophy* that 'We never discover a train of contrivance to bring about an evil purpose' or 'to produce pain and disease'. Paley's attention rests almost exclusively on traits that show a utilitarian contribution towards the survival of individuals, or otherwise makes their lives easier. Paley (1785, p. 42) provides an early theological expression of utilitarian ethics where adaptation is evidence that 'God wills and wishes the happiness of his creatures'. Paley (1802) went on to use such reasoning as the basis of *Natural Theology*, where he uses the appearance of design to make a case for a benevolent Creator. In *The Origin of Species*, Darwin's theory of evolution by natural selection naturalised Paley's argument, switching the cause of adaptation from a supernatural Creator to the natural environment. In doing so, Darwin (1859, p. 211)

explicitly imports the idea that the designing agent—natural selection—
'acts solely by and for the good of each'. In making this substitution,
Darwin has to switch the goal of adaptation from the happiness of individual organisms to increasing their ability to survive and reproduce. For
all that this is a subversion (Gould 2002, p. 127), Darwin maintains the
essentially benevolent character of the designer—even if that design now
comes from natural selection rather than the direct handiwork of God.
Without accepting the benevolence of natural selection, Darwin would
not labour to repeat throughout *The Origin of Species* that natural selection acts to the benefit of individual organisms (1859, pp. 109, 125, 149,
172, 233, 471, 489, etc.).

There is one potential point of departure from this narrative, which
concerns the discussion of sexual selection. Darwin (1859, p. 88) argues
that

> Generally, the most vigorous males, those which are best fitted for their
> places in nature, will leave most progeny. But in many cases, victory will
> depend not on general vigour, but on having special weapons, confined
> to the male sex. A hornless stag or spurless cock would have a poor chance
> of leaving offspring.

Here, Darwin does seemingly separate fitness or vigour from producing
offspring. But Darwin (1859, p. 127) goes on to extend his understanding of adaptation to include these traits, in that 'sexual selection will give
its aid to ordinary selection, by assuring to the most vigorous and best
adapted males the greatest number of offspring. Sexual selection will also
give characters useful to the males alone, in their struggles with other
males'. In this way, Darwin (1859, p. 158) uses adaptation to encom-
pass traits that help in the full range of an organism's conditions of life,
as if 'to fit the two sexes of the same species to each other, or to fit the
males and females to different habits of life, or the males to struggle with
other males for the possession of the females'. As such, horns, spurs, and
other special weapons are adaptations in that they support the ability
of an individual male to reproduce, which Paley does not address. In
what could be construed as a rare moment of moralising, Darwin (1859,
p. 472) does reflect that some of these traits are 'abhorrent to our ideas

of fitness', but this simply reflects how Darwin's perspective is borne out of the philosophical influence of Paley's utilitarianism.

For Darwin, the concept of adaptation reflects the fit between an individual organism and the design specification of survival and reproduction in the environment. So, like an archer aiming for a bullseye, there is an absolute standard of success in producing more offspring. Darwin (1859, p. 66) also urges that

> it is most necessary to keep the foregoing considerations always in mind— never to forget that every single organic being around us may be said to be striving to the utmost to increase in numbers; that each lives by a struggle at some period of its life; that heavy destruction inevitably falls . . . during each generation or at recurrent intervals.

So, as when winning for an archer depends on shooting closer to the bullseye, the number of those offspring that go on to produce offspring of their own depends on how successful others are in producing offspring, where Darwin (1859, p. 468) argues that the 'better adaptation in however slight a degree to the surrounding physical conditions, will turn the balance'. In brief, this is Darwin's resolution of ecology and evolution. Regardless of the intensity of competition, Darwin (1859, p. 79) establishes the common-sense notion that 'the vigorous, the healthy, and the happy survive and multiply', so the gains of adaptation should generally lead to an increase in population. In Darwin's understanding of how natural selection works through the competition of individuals, there is no room for an equal and opposite counterpart for adaptation in maladaptation; an organism that is less well-suited to its environment is like an archer that consistently shoots further from the bullseye: it will lose the competition.

Darwin's approach to competition was brought into the foundations of modern evolutionary biology in population genetics by Fisher (1930, pp. 22–34) through his rigorous mathematical treatment of fitness as lifetime reproductive success. Fisher demonstrated how to measure fitness by adding up all the offspring that an individual has over its lifetime, which provides a practical method for measuring adaptation to study evolution by natural selection. At any given point in time, it has an absolute value in terms of the number of offspring that are produced by an individual. Strictly, it would be necessary to measure the number of offspring that survive to the same point in the life cycle that is under measurement, which is practically challenging. Fitness depends upon all of the traits that an organism has, which reflects its overall quality, which Fisher (1930, p. 38), following Darwin, referred to as adaptation:

> An organism is regarded as adapted to a particular situation, or to the totality of situations which constitute its environment, only in so far as we can imagine an assemblage of slightly different situations, or environments, to which the animal would on the whole be less well adapted; and equally only in so far as we can imagine an assemblage of slightly different organic forms, which would be less well adapted to that environment.

Fisher also partitions adaptation, as measured by fitness, into the average effect that different genes have on the variation in fitness that is attributed to genetic differences within the population. This can be partitioned all the way down to the average fitness of individuals with a mutant or wild-type gene at a specific locus within the genome. In his fundamental theorem, Fisher (1930, p. 35) shows that, from a population perspective, 'The rate of increase in fitness of any organism at any time is equal to its genetic variance in fitness at that time'. And so, by treating fitness as the currency of adaptation, Fisher establishes a logical relationship between natural selection changing the genetic constitution of the population by increasing fitness and likewise adaptation. It is therefore unsurprising when Fisher (1930, p. 47) goes on to explicitly suggest that, from greater adaptation, 'Any net advantage gained by an organism will be conserved in the form of an increase in population'.

Fisher's simplistic treatment of the connections between adaptation, fitness, and population was not unchallenged. Haldane (1932, p. 126), building on the logic of Peter Kropotkin (1902) and hinting at his own political leanings, suggested that 'the special adaptations favoured by interspecific competition divert a certain amount of energy from other functions, just as armaments, subsidies, and tariffs, the organs of international competition, absorb a proportion of the national wealth which many believe might be better employed'; Wright (1945, p. 417) made a similar argument. Huxley (1942, p. 485), also hinting at his own ethical leanings, went on to say it could no longer be claimed that 'the more ruthless the competition, the more efficacious the selection, and accordingly the better the results' because 'we now realize that the results of selection are by no means necessarily "good", from the point of view either of the species or of the progressive evolution of life. They may be neutral, they may be a dangerous balance of useful and harmful, or they may be definitely deleterious'. Despite how these quotes might be first appear, Haldane and Huxley are merely making a moral protest about the costs of competition to a species (that echo Darwin's own moral sentiments), which was commonplace around the time of the World Wars; there was never any suggestion that natural selection leads to outcomes other than adaptation, but rather that adaptation was not as 'morally good' as it might at first seem. It is therefore unsurprising that, regardless, Fisher's reasoning was later consolidated into orthodoxy by Williams (1966, p. 25) in the claim that 'Natural selection would produce or maintain adaptation as a matter of definition'. So, Darwin's (1859, p. 201) dictum that 'natural selection acts solely by and for the good of each' persisted, even if traits that help the individual organism can harm the species or life as a whole. In this account, there is no possibility of maladaptation that is naturally selected because it harms the individual organism.

But there is a problem in Fisher's reasoning, which must be correctly ascertained. Many people have taken aim at the simplistic treatment that Fisher (1930, p. 47) offers between increased fitness and increased population because of specific phenomena (e.g. Allee effects) that complicate matters. However, Fisher, in setting out a general perspective, is only meaning to support Darwin's common-sense notion that, all else being equal, greater adaptation through an increased fit between an organism and its environment should enable the species to increase in population. If this is not generally the case, then many phenomena in evolutionary ecology (e.g. adaptive radiations) become a much bigger problem to explain. Instead, the right complaint is that two distinct meanings of fitness have come into use in ecology and evolution. There is the absolute fitness of individual organisms as lifetime reproductive success in units of offspring, and there is the relative fitness of alleles in units of their differential contributions to lifetime reproductive success in arbitrary units. For an allele to be favoured by natural selection, its needs to have higher relative fitness than competing alleles at the same locus. Indeed, the arbitrariness of the units is emphasised by the fact that it is common practice in population genetics to assign a wild type a relative fitness of 1, whilst any mutant has a fitness that is expressed as a deviation from the wild type using a selection coefficient (e.g. $1 + s$). So whilst absolute fitness corresponds to Darwin's concept of the adaptation of individual organisms, relative fitness is a genetic concept. All else being equal, Fisher (1930, pp. 34–35) only mathematically showed that natural selection increases relative fitness; he did not show that natural selection exclusively increases absolute fitness. Consequently, there is room to distinguish between archery and boxing for how a gene contributes towards an individual's survival and reproduction. Evolution by natural selection can lead to two resulting types of traits: adaptations that increase absolute fitness and maladaptations that decrease absolute fitness.

Others have also recognised that there are two distinct meanings of fitness. Although it has often been recognised without further consideration or suggested to be of little importance (e.g. Ghiselin 1974, pp. 49–51; Sober 1984, pp. 15–17), the two meanings of fitness have recently come to the fore in the study of eco-evolutionary dynamics. As Graham Bell (2017, p. 606) makes clear in his review of evolutionary rescue,

> The foundations of ecology and evolution rest on two different metrics. In ecology, the fundamental parameter is the rate of increase of a population, which governs its abundance. In evolution, this is replaced by the rate of increase of a type within a population relative to the weighted average of all types, which governs the change in composition. This amounts to a distinction between absolute fitness, as used in ecology, and relative fitness, as used in evolution. The basic innovation of eco-evolutionary dynamics is to incorporate both absolute and relative fitness within the same framework.

So recognising a distinction between absolute and relative fitness is nothing new.

But, interestingly, Bell does not open wide the difference when it comes to understanding the possibilities for evolution by natural selection in imagining a maladaptive trait that increases relative fitness at the expense of absolute fitness. Instead, Bell (2017, p. 606) reasserts that

> Heritable variation in relative fitness will necessarily tend to alter the mean phenotype of a population, but this will lead to permanent adaptation only if the absolute fitness of the type with the greatest relative fitness is positive. This distinction is particularly important when the environment deteriorates to such an extent that a population, as presently constituted, is unable to replace itself and must therefore dwindle over time until it becomes extinct.

Bell, and also others working on eco-evolutionary dynamics (see Hendry 2016), follow Fisher's (1930, pp. 42–46) reasoning that an absolute increase in fitness is required for evolution by natural selection. A proposal of something like maladaptation would appear to be associated with population extinction, which is a slippery-slope argument when a maladaptation may simply decrease population size without bringing it close to collapse. Context may be important to understanding why Bell and others hold this view: in focusing on evolutionary rescue, the interest is in adaptations that increase absolute fitness so that a population can persist in a changed environment. As such, perhaps the neglected possibility for maladaptation is unsurprising, and maybe this is where the concept could be most readily taken up to further existing theory. Nonetheless, even in this most closely related field of study, researchers have not hit upon the concept of maladaptation, and so it would seem that it would still be viewed as impossible. Long is the legacy of Paley's natural theology in evolutionary biology.

* * *

In the potential friction between the logic for maladaptation's possibility and the historical bias against it, it is worth being exceptionally clear about why the proposed concept of maladaptation is needed. Expert readers may well be thinking that the case that is put forward for the concept of maladaptation is either obvious or false. This dichotomy comes up time and again in the sciences, which is to be expected when, in an aphorism attributed to Thomas Huxley, science is taken to be 'trained and organised common-sense'. The fallout of this position is seen in Huxley's own intellectual journey, when he gave a ferocious review of the early evolutionary ideas in Robert Chambers' (1844) *Vestiges of the Natural History of Creation*, but later, when asked about the argument for natural selection in *The Origin of Species*, is famously reported to have replied that he felt 'extremely stupid not to have thought of that'. As mere common sense, perhaps there is a narrative presentation of the concept of maladaptation that would suggest that it is perfectly consistent with current theory, and thereby 'obvious'. To construct all new ideas in the accommodation of old ones would certainly make for an easier path to acceptance by fellow scientists, but it can also be intellectually dishonest, if not downright unscientific. An existing theory that is true should be able to be consistent with another true theory that is about something else, but if an existing theory can accommodate a relevant new observation that it did not previously explain then it would appear to have the cowardly flexibility of pseudoscience—rather than being formed by 'the [scientific] method of bold conjectures and ingenious and severe attempts to refute them' (Popper 1972, p. 81). In this regard, the growth of scientific knowledge is like the Herculean fight against the hydra: if a theory fails, it is cut off and new theories will then grow to more than take its place.

In the case of maladaptation, existing concepts have not, up until now, satisfactorily suggested that natural selection would produce the full implications of the traits that are here described as maladaptations. This is not least because maladaptation cannot just 'happen'; it only arises because of the ecological effect, which is why it has not fallen out of widely used mathematical models in evolutionary biology. Clearly, the fundamental error has been assuming a correspondence between natural selection simultaneously increasing absolute fitness and relative fitness. A large part of this error stems from the lack of integration of ideas from evolution and ecology, where maladaptation is a paragon of the changes that need to be made to evolutionary theory. In search of greater unity, the theory underlying the error must be cleaved from the body of work into evolution by natural selection.

Accommodation of maladaptation as a part of existing theory would require a substantial rewrite of history. Some may prefer to accommodate what are referred to as maladaptations as a special kind of adaptation; for example, following Williams (1966, p. 25), as a holistic description of the fit between an organism and its complex environment—whether like archery or boxing. Against this stance, maladaptation is arguably an apt term, literally meaning a bad adaptation, which Paley would be forced to cast as doing evil to an individual organism. Maladaptation still represents adaptation in the sense of a fit between an organism and its environment, but it is bad in the sense that it contravenes the dictum that 'natural selection acts solely by and for the good of each' (Darwin 1859, p. 211). Why is this significant? By harming an individual's ability to survive and reproduce, a maladaptation contravenes the benevolence of natural selection. If natural selection can lead to maladaptation, it is no longer tenable to support Paley's (1785, p. 40) assertion that 'all the contrivances which we are acquainted with, are directed to beneficial purposes'. For this reason, maladaptation represents a departure from—not an extension to—the foundations of evolutionary biology. There are times for the defence of existing ideas to avoid wasting breath on confusion, and there are times for innovation to open up new problems for researchers, and this is the latter because the goal of the theory of evolution by natural selection becomes more than 'to explain the same set of facts that Paley used as evidence of a Creator' (Maynard Smith 1958, p. 82) in also explaining what Paley could not. Indeed, maladaptation would perhaps only be one part of the open opportunity for a renewed field of study: to explore the population biology of natural selection in the interplay of evolution, genetics, and ecology.

If opinions do divide along these lines, it would reveal something important about the perspective from which evolutionary change is viewed. The new concept of maladaptation takes a designer's perspective, which may be contrasted with a designee's (i.e. individual organism's) perspective. Adaptation can be understood in a unitarian way via a degree of flexibility in its definition, whereupon natural selection can be said to favour traits that contribute to increasing an individual organism's survival and reproduction in some contexts and not in others. This may make sense when adaptation is intended to refer to the fit between an organism and all aspects of their environment. Such an approach may be a reasonable use of terms because, to an individual, it is not especially relevant if their traits lead to their survival and reproduction increasing or decreasing; with analogy to long-distance running, this is much like how the incentives of a competition are only relevant to a runner in as far as they know how to follow them. Alternatively, making room for maladaptation involves splitting the unitarian concept of adaptation into a newly defined adaptation and maladaptation. This may make sense when adaptation is intended to refer to the fit between an organism and a benevolent designer's specifications. Such an approach, following from Darwin and his interest in artificial selection, may be a reasonable use of terms because, to a designer, it is highly relevant whether or not a benevolent outcome is obtained from the incentives that are intentionally provided. Regardless of which approach is clearer, the designer's perspective is the heritage of evolutionary biology and the conceptual world-view that has already been bought into (cf. Gould 2002, p. 158). And perhaps it is for good reason, because it is a powerful construction of theory that quite literally takes a God's-eye view of nature. This construction may be most helpful when trying to use the theory to shape the direction of evolutionary change through artificial selection, whether in domesticated stocks, laboratory experiments, or field studies.

The conceptual construction of a theory has a deep-seated impact on its development, and rarely leaves its devotees free to think from its biases. Perhaps the most pressing reason why the concept of maladaptation is needed is to set scientific imaginations free. By breaking the basic association between natural selection and adaptation, there is no reason to think that natural selection would generally lead to an increase in population. If natural selection can favour traits that harm individual organisms, the intrinsic evolutionary trajectory of life is no longer assuredly on the path of increasing adaptive efficiency, diversity, and complexity because it could equally move (of its own accord) towards a maladaptive catastrophe in extinction. This is not to suggest that the connection between maladaptation and extinction is trivial—just as the connection between adaptation and efficiency, diversity, or complexity is not trivial—but it does shake the general conception of what evolution by natural selection does. It is in this vernacular sense that maladaptation is the product of evolution by natural selection in what would previously have been viewed as the 'wrong' direction.

Beyond opening the door to the extreme possibility of extinction through evolution by natural selection, for a more moderate role of maladaptation, there is an intellectual analogy. In economics, the concept of 'market failure' or externalities (*sensu* Pigou 1920) serves a similar role to that which maladaptation could serve in evolutionary biology. In classical economics, market competition drives down the prices for goods by encouraging innovations including technological breakthroughs, better firm organisation, and other changes that make the production process cheaper. Markets are therefore viewed as an efficient vehicle for allocating resources within an economy, giving the greatest profits to those that are best placed to reinvest them. Reality is often very far from the ideal picture that classical economics presents (even setting aside illegal activity), and so neoclassical economics recognises that inefficiency can occur with market failure. Recognising this possibility has led to the study of the causes of market failure, which often link back to monopolism, externalities, and/or public goods. Once the causes have been understood, there is then the possibility of developing solutions like anti-trust law, Pigouvian taxes, and regulatory standards to overcome these market failures. As such, the admission of the imperfection of markets by economists has proved to be a highly useful theoretical innovation. Market failure reinforces that although the prevailing direction of change may bring about a benevolent outcome in idealised theory, there are predictable reasons that regularly bring about malevolent outcomes in reality. The situation is very similar to maladaptation: foundational theory in evolutionary biology to date has often been based on mathematical models that assume unstructured populations with random mating (etc.), which is indisputably unrealistic, and now the phenomenon of maladaptation shows that violating these assumptions can make natural selection produce a qualitatively different kind of outcome, as natural selection can decrease absolute fitness and even drive extinction.

Adopting the concept maladaptation could start the recognition of a kind of 'evolutionary failure' (*contra sensu* Bradshaw 1991) in biology, which could lead on to important developments. As a tangible case in point, there has been substantial attention focused on the development of genetic technologies for the control and eradication of populations or species that are a threat to human activities (e.g. Price et al. 2020). Objectives are wide-ranging, including the eradication of mosquitoes that vector diseases such as malaria, dengue, and zika, and the conservation of communities against invasive species that are destabilising ecosystems across the world. A major problem for these technologies is the evolution of resistance. If selection pressures could be presented in such a way as to promote maladaptation, then this may provide insights into preventing resistance to these new technologies, and so make them evolutionarily robust (e.g. Madgwick and Wolf 2021). Alternatively, maladaptations could be engineered away, particularly those that are discovered to have a role in human diseases. There is much more to study here.

In the end, then, it is useful to adopt the redefined concept of maladaptation on the basis that the theory of evolution should be set up in a way that best empowers people to use it, but there is no need to be prescriptive in suggesting that the redefinition must be accepted. The redefinition is taken for granted here, so that the term can be freely used. But, it must be made clear, this book is not principally about advocating the use of the term maladaptation. It is about the phenomenon that the term describes: those traits that decrease individual survival and reproduction, whatever they are called. And so, having outlined the reasons for the possibility of maladaptation, the next chapter moves on to discuss the challenges of uncovering what are perhaps its most obvious examples amongst the social behaviours of individual organisms.

4

Maladaptation in social behaviour

Maladaptations in the social behaviours of individual organisms may seem like the most obvious context from the theory of maladaptation that has been outlined. To be convincing, each example needs to assemble numerous lines of evidence. Consider the cannibalistic behaviour of the flour beetle *Tribolium confusum*, which has been a long-running model system for research into evolution and ecology (Pointer et al. 2021). Flour beetles engage in both egg and pupa cannibalism in laboratory settings that mimic the semi-natural environment of flour stores (Alabi et al. 2008). Individuals eat both their own and their neighbours' offspring, which is harmful to everyone. There is heritable variation in the cannibalistic tendency that is attributable to allelic variation at a small number of genetic loci (Stevens 1989). Varying rates of cannibalism across strains lead to large differences in population sizes in the laboratory after just one generation (i.e. due to reduced absolute fitness), but can also rapidly lead to large differences in the population sizes under controlled conditions over time—and have even been known to drive laboratory populations to extinction (Park et al. 1964). This apparent harmfulness reveals the potential for cannibalism to be maladaptive, even if extinction is simply the result of non-adaptation to the laboratory conditions. Beyond potential maladaptiveness, experimental evolution can lead to higher rates of cannibalism (and lower population sizes) but only when there is local competition for mates within the group (Wade 1980), which is more suggestive of maladaptation. Yet, there are many adaptive explanations of this behaviour, including buffering against resource poverty

(Fox 1975), and it may be that egg and pupa cannibalism evolves for different reasons (Alabi et al. 2008). But the genetic evidence suggests that variation in cannibalism on eggs and pupae is controlled by the same loci (Stevens 1989). Therefore, as it stands, this example both seems to fit the definition of maladaptation and shows some of its key hallmarks. It is possible to criticise the attribution, especially on the basis that the laboratory is not the same as nature; even the experimental evolution only shows that it is possible for the environment to promote maladaptive cannibalism, but it cannot show that this is the rationale for cannibalism in nature (without further demonstrating that the ecology is the same). It is very challenging to find conclusive evidence of the properties that are critical to maladaptation.

For many potential examples of maladaptation, it remains very hard to obtain appropriate evidence. A consistent problem in seeking evidence from most evolutionary model systems is that they neglect to explore the ecological impact of trait evolution on population size, which is a general frustration in retrospectively applying the new concept of maladaptation to the results of experiments that were not intended for such use. Herein, there can be a temptation to seek examples of maladaptation among predefined types of social behaviour. But maladaptation does not correspond to any of these. Consequently, there are three common and distinct pitfalls that are encountered, with the misguided inference that examples of intraspecific competition, spite, and genetic conflict can be presumed to imply maladaptation. In each case, the absolute fitness that is used to define maladaptation is mistakenly interpreted to be the focus of the classification of these types of social behaviour, rather than providing a new dimension to their classification. Nonetheless, these phenomena may supply examples of maladaptations on a case-by-case basis, which must be carefully considered.

The first pitfall comes from associating maladaptation with all traits involved in intraspecific (or interference) competition, which enhance the competition among members of the same species. As previously discussed, evolutionary biologists such as Kropotkin (1902), Haldane (1932), and Huxley (1942) have discussed intraspecific competition as a kind of wastefulness, especially in application to human warfare. Later writers were much less sceptical about the value of competition. Whilst competition is good for some individuals and bad for others, it is a way of allocating resources that ideally results in giving them to those that are best placed to use them. This is ultimately the basis of the Darwin's 'struggle for existence':

> As natural selection acts solely by the preservation of profitable modifications, each new form will tend in a fully-stocked country to take the place of, and finally to exterminate, its own less improved parent or other less-favoured forms with which it comes into competition. Thus extinction and natural selection will, as we have seen, go hand in hand. (Darwin 1859, p. 172.)

Whilst it can be carried out in ways that are morally unpleasant, competition within the species can be the driver of evolutionary innovation that increases absolute fitness.

Yet, even some of its fiercest champions like Matt Ridley (1996, p. 34) have sometimes oversimplified the relationship to assume that intraspecific competition is undesirable for the species; for example, 'Eaglets often kill their younger brothers and sisters in the nest. Good for the individual, bad for the species'. If it is accepted that siblicide is bad for the species in the sense that fewer eaglets are reared to adulthood, it must also be reflected that it could be good for the species in the number of eaglets that are reared in the next generation. Although it is morally unpleasant, with the strongest eaglets killing off the weakest, the strongest eaglets obtain more resources for themselves, which can enhance the probability that they survive to raise their own offspring. Indeed, it is common for many raptor parents to lay multiple eggs as an insurance policy, even though they only have the resources to raise one chick to adulthood (Morandini and Ferrer 2015); if their first-born is a healthy-enough offspring then the parents do not interfere as it kills off its siblings, whereas if it is sickly or still-born then it is liable to be supplanted by a younger sibling. Perhaps Ridley (1996, p. 34) is right that such competition could harm the survival of a rare eagle species from the point of view of a human conservationist, but it is certainly debateable whether one well-fed offspring is a better contribution to the species with siblicide than many malnourished offspring without siblicide. So even though siblicide is a harmful trait, it is still likely to be an adaptation, albeit a gruesome one—and it may even non-adaptively drive the species to extinction as the environment changes.

In distinguishing the sort of competition that increases survival and reproduction of individuals (and species) from the ruthless competition that generates maladaptation, there are many surprising adaptations. Handicap signalling is an obvious example that has been described in almost maladaptive terms due to its costliness (Zahavi 1975). The classic example of a peacock's train must be extremely costly for males to carry, making flying away from predators much more difficult. Yet handicap theory would suggest that the costliness is functionally essential for the honest signalling of quality by competing males (Zahavi and Zahavi 1997, p. 40):

> Natural selection encompasses two different, and often opposing, processes. One kind of selection favors straightforward efficiency, and it works in all areas except signaling. This selection makes features—other than signals—more effective and less costly; we suggest calling it 'utilitarian selection'. The other kind of selection, by which signals evolve, results in costly features and traits that look like 'waste'. It is precisely this costliness, the signaler's investment in the signals, that makes signals reliable. We suggest calling this process 'signal selection'.

Peacocks with more elaborate trains reduce their survival to enhance their reproduction. In this way, although there is a costliness associated with the more elaborate train, the handicap of a peacock's train might be an adaptation.

The extreme elaboration of the peacock's train could suggest that the intensity of male–male competition has tipped the balance of the trait into maladaptation, but this is not necessarily the case. The key point is that raising the bar for competition is not enough to infer maladaptation. The success of a male peacock depends on how its train compares to other males. Consequently, it is possible to imagine an ancestral peacock with a much more modest train that could have been more successful than a contemporary peacock now is despite its elaborate train, because the bar for success in the train competition has been raised over evolution by natural selection, like in an arms race. The allocation of more resources to male–male competition over time is not maladaptive, but merely reflects the change in the environment, which here is determined by the trains of other males (Huxley 1942, pp. 438–441). In this sense, there is an observable fit between what a peacock does and what the environment demands to increase its survival and reproduction, which might suggest that it is still an adaptation—albeit a surprising one.

Sometimes, raising the bar for success in a competition can have more complicated results, such as in classic examples of the 'tragedy of the commons' (Hardin 1968; see also Rankin et al. 2007). For instance, in the slime mould *Dictyostelium discoideum*, under an environmental stress like starvation, free-living cells aggregate together to form a multicellular fruiting body to hold spores aloft to aid their dispersal to find more food (Strassmann et al. 2000). In the fruiting body, the cells take on specialised roles, differentiating into reproductive spore and nonreproductive stalk cells, which gives the potential for conflict. Further, as cells often have a recent common ancestor from rounds of asexual reproduction in a location, a small number of cell strains can aggregate to act out the potential conflict. When different strains form a fruiting body together, they can adjust their allocation of spore and stalk cells to cheat the other strain(s) by producing more 'selfish' spore cells that go on to reproduce, and fewer 'altruistic' stalk cells that die in the fruiting body (Madgwick et al. 2018). When one strain cheats another, it produces more dispersing spores; so conditional cheating would appear adaptive. Consequently, conditional cheating is expected to spread through the population, so that multiple strains acquire this competitive trait. When both strains try to cheat each other, they both produce no or fewer dispersing spores because too few stalk cells are produced (Belcher et al. 2022); so, now that all strains are doing it, conditional cheating would appear maladaptive. Of course, from the perspective of a cheated strain, it is better to produce fewer stalk cells to cheat the other strain even if the other strain is cheating them, because the bar for competition has been raised. So each step in the evolution of the trait is favoured by natural selection, but the end result is maladaptive. There is no problem in changing perspective on whether a trait like spore/stalk cell allocation is adaptive or maladaptive depending on its fit with its environment (of other strains' spore/stalk cell allocations), but perhaps it is not a clear example of maladaptation as a result. Such frequency dependency may sound like an unusual situation, but its general features recur wherever there is a public good-like trait that can be exploited by cheating.

Public goods problems that result in a tragedy of the commons are widespread occurrences across nature (Rankin et al. 2007), from competition among viruses over the use of a host's or each other's resources (Leeks et al. 2021), through the production of iron-scavenging molecules in bacteria (West and Buckling 2003), to the production of body heat in vertebrate nests (Haig 2010). The existence of these failures should indicate to us that it is not always possible to do something about them. Yet, unsurprisingly, their occurrence can create selection for avoidance mechanisms. For example with *D. discoideum*, strains have been shown to segregate out of a fruiting body in groups that are likely to end in failure through fruiting-body collapse before dispersal can take place, due to too few stalk cells being allocated by the group (Madgwick et al. 2018; Belcher et al. 2022). In other systems, similar results can be obtained by adaptations in a powerful individual to takeover or coerce cooperation, such as a parent producing body heat for their offspring (Haig 2010) or a dominant female meerkat preventing subordinate reproduction to get more help in raising their own offspring (Clutton-Brock et al. 2001). However, similar results can also be obtained by maladaptations, as one solution is for an individual to monopolise a resource that could be mismanaged by a group, whereupon the incentives to efficiently use a superabundant resource can vanish (Knowlton and Parker 1979). So resolutions should not be assumed to be adaptive. This means that it can be difficult to ascertain if a trait is a maladaptation at first glance because, as attempts at resolving public goods problems show, the rules of the competition are also subject to evolutionary change.

Therefore, to avoid the first pitfall, it must be recognised that intraspecific competition is insufficient to infer maladaptation. Whilst competition creates winners and losers by allocating resources to those that can use them best, it is possible that on average individuals tend to use resources more efficiently. Maladaptation only arises when competition becomes more intense and ruthless, where individuals can benefit from lowering absolute fitness to increase their relative fitness. On a case-by-case basis, there are some traits involved in intraspecific competition that are appear maladaptive, which can be progressively illustrated out of some cases of sexual conflict, where one sex harms the other during reproduction.

In the bean weevil *Callosobruchus maculatus*, male genitalia have spines that leave conspicuous scarring during copulation with females (Hotzy and Arnqvist 2009). The longer-term costs of the harm that males inflict on females include reduced fecundity and longevity. The harm to females in of itself did not improve male fitness in comparisons where the harm caused was constant. Instead, males appear to be favoured by natural selection for harming females because it enhances their success in sperm competition against other males, which can be shown by the positive correlation between genital spine length and fertilisation success. Nonetheless, the harm to females is not strictly a byproduct because the harm to females that they mate with is also a harm to males' reproductive success, and so the harm to females would seem to be essential to the evolution of enhanced sperm competitiveness in males. Consequently, in this example, it is difficult to argue that the competition is merely distributing resources to those that can use them best because male bean weevils are trashing the prize they are competing for (i.e. offspring from mating with females).

A more quantitative demonstration of the capacity for sexual conflict is found in the model fruit fly *Drosophila melanogaster* (Holland and Rice 1999). Female fruit flies typically mate with multiple males, which can lead natural selection to favour female-harming traits in males. The costs of such harm to females can be difficult to detect in natural populations. But in a laboratory, the costs can be brought out by evolving populations that are forced to be monogamous, with one male mating with one female, which gives males and females aligned interests. After 47 generations, controlled matings between males and females can be used to compare the derived and ancestral lines to quantify the cost of sexual conflict in natural populations. Derived females had responded to the artificial environment by evolving reduced resistance to male courtship and seminal fluid. Derived males were also found to have evolved to be less harmful to the reproductive output of their mates. Overall, monogamy increased net reproductive value, leading to an estimated load from sexual conflict in the natural population of close to a 20% decrease in offspring produced.

A final, more extreme example of sexual conflict comes from male aggression in the common lizard *Lacerta vivipara* (Le Galliard et al. 2005). With the experimental manipulation of the sex ratio in natural populations, male-biased populations experience enhanced aggression towards adult females, resulting in lower female survival, fecundity, and emigration rates. The effects on females can be quantitatively tied to male behaviour, as harm to females can be demonstrated through having close to three times the number of mating scars and back injuries in male-biased populations compared to female-biased populations. Male aggression generates a positive feedback cycle, whereby increased aggression in male-biased populations contributes towards more male-biased populations. Unsurprisingly, such populations decline in size and have been noted to have an increased probability of extinction.

* * *

The second pitfall arises from equating maladaptation and spite, which really stems from a misunderstanding of the relationship between maladaptation and inclusive fitness. Maladaptation has been described in terms of decreasing individual fitness, making individuals worse at survival and reproduction. This becomes tricky to interpret for social behaviours because an individual's fitness is interdependent with what is going on in their social environment. Inclusive fitness is a quantity that describes how natural selection acts when a trait of an actor affects the fitness of a recipient, resolving the fitness that is attributable to an individual and their traits regardless of who gets the fitness effects. This is not the same as 'individual fitness' in its classic definition as 'lifetime reproductive success' (*sensu* Fisher 1930) because the classic calculation strips away the (presumed weak) effects that individuals have on each other's fitness (Marshall 2015, pp. 56–57). But it is individual fitness when properly accounting for social effects.

Inclusive fitness was developed for the study of altruism (Hamilton 1964b). Canonical examples of altruism include sterile workers in social insects (Hamilton 1972), helpers at the nest in cooperatively breeding birds (Hatchwell 2009) and cell-sacrifice in fruiting body formation in slime moulds (Strassmann et al. 2000). To be favoured by natural selection, such traits need to generate positive inclusive fitness through the costs of altruism to the actor being outweighed by the benefits to a recipient for the causal gene(s). For most examples (West et al. 2007b), the positive relatedness that tips the balance in favour of altruism arises due to interactions occurring among kin, who have a probability of sharing genes by common ancestry. Although there can be a role for genetic recognition (Bourke 2011, pp. 130–137), the mechanism that generates positive relatedness is usually limited dispersal (West and Gardner 2013).

Spite is a type of social behaviour that is near enough the opposite of altruism, describing traits where an actor pays a cost to harm a recipient (Hamilton 1970). Just as where altruism that helps a recipient that is positively related generates positive inclusive fitness, spite can generate positive inclusive fitness through giving harm to a recipient that is negatively related. Negative relatedness may sound paradoxical from the perspective of reflecting a probability of sharing genes, but the coefficient of relatedness that is important for social behaviour is the probability of sharing genes relative to a random expectation (Grafen 1985). Accordingly, negative relatedness arises when an individual is less likely to share genes than the random expectation. Unsurprisingly, despite its theoretical possibility, it is difficult to imagine how this could occur, and so for a long time spite was 'Hamilton's unproven theory' (Foster et al. 2001, p. 229). These difficulties were obvious from the start; when Hamilton

(1970, pp. 1219–1220) originally discussed several potential examples, in each case he concluded that they were unlikely to be spite for three reasons.

Hamilton's (1970, p. 1220) first reason was that 'all actions do cost something', which is a little cryptic but becomes comprehensible with an understanding that costly spite may not be favoured by natural selection when it is rare because its costs would outweigh its benefits. This was elaborated in Madgwick (2020), which clearly showed how negative relatedness based on any kind of mechanism that negatively assorts self and nonself alleles is frequency dependent. When a spiteful allele is rare, it is expected to have near-zero negative relatedness, which increases with its initial increase in allele frequency (see later). An allele is favoured by natural selection based on its mean fitness across interactions. A rare allele for spite only interacts with one (or a few) nonself allele(s) and so, whilst the rare allele for spite has its mean fitness largely determined by the costs of spite, the common allele mostly has its mean fitness determined by its non-spiteful interactions with itself. As such, no matter the effect of the harm on the recipient, a rare allele for spite just cannot interact with enough competing alleles to outweigh the cost to itself. By contrast, cost-free spite only has benefits and so would always be favoured by natural selection, and on this basis cost-free spite is not theoretically similar to spite (Keller et al. 1993). Moreover, in keeping with Hamilton's first point, cost-free traits are probably implausible in nature.

Hamilton's (1970, p. 1220) second reason was that 'an animal will not normally have any way of recognizing which other members of its species have less than average relationships'. Indeed, at the time, genetic recognition seemed far-fetched (Dawkins 1976). But Hamilton's argument (as he elaborates) only applies to populations that have positive assortment by genetically blind processes that arise from the migration and movement of individuals—such as limited dispersal—that increase the likelihood of neighbours sharing genes. Whilst these alter the random expectation of sharing genes within a neighbourhood, it does not differentiate the positively and negatively related from one another in a way that could be acted upon. Moreover, as Hamilton (1971) discussed elsewhere, populations do not generally have negative assortment. One conceivable exception that was discussed occurs when an individual may be favoured by natural selection to seek out an unrelated mate. This practice of disassortative mating is common enough in nature to avoid inbreeding costs, but mates tend to have a shared interest in the reproduction of offspring, which might remove the scope of positive inclusive fitness from spite (though refer back to the examples of sexual conflict).

Lastly, and more interestingly, Hamilton's (1970, p. 1220) final reason was that 'single populations that are so small in "effective size" as to have [negative relatedness] much different from zero must be in a precarious situation already, and the selection of a gene causing spite can only hasten their extinction'. Hamilton makes a connection between spite and population extinction, which suggests that he was thinking of spite as a trait that decreases survival and reproduction. Whilst this provides a first tentative link between maladaptation and spite, it is not one that is widely agreed upon. The original meaning of spite from Hamilton (1970) has been systematically reinterpreted within the subsequent literature. Spite, in its modern understanding, is explicitly referred to as a 'social adaptation' that increases individual success (see e.g. West et al. 2007a; Bourke 2011, pp. 28–30; Foster 2011; Marshall 2015, pp. 25–28). Indeed, in the most recent dedicated review (Gardner and West 2004), the connection between spite and population extinction is scarcely mentioned—and the passing reference is made only as a point of disagreement with Hamilton.

Why has this happened? The vast majority of research that came after Hamilton (1970) was done by researchers who followed Williams' (1966, p. 25) argument that 'Natural selection would produce or maintain adaptation as a matter of definition', which ultimately harks back to Darwin's (1859, p. 211) maxim that 'Natural selection will never produce in a being any structure more injurious than beneficial to that being, for natural selection acts solely by and for the good of each'. Consequently, for most evolutionary biologists, there was little room for the idea of a trait that genuinely harms individual fitness. The first major step in the adaptive reformulation of the concept of spite came just a few years after Hamilton (1970) in the publication of the influential textbook

Sociobiology (Wilson 1975b). Hamilton (1970) defined spite with respect to an actor and a recipient, where the actor pays a cost to harm the recipient. In its interpretation, Edward Wilson (1975b, p. 119) made sense of spite being favoured by natural selection by introducing a third party (or secondary recipient) that benefits from the harm: 'the spiteful individual lowers the fitness of an unrelated competitor . . . while reducing that of his own or at least not improving it; however, the act increases the fitness of the brother to a degree that more than compensates'. In other words, Wilson (1975b, p. 119) reconceptualises the harmfulness of a spiteful interaction as a benefit to an individual that is not part of the interaction, as a form of indirect altruism.

Interestingly Wilson (1975b, p. 119), much like Hamilton (1970), maintained that 'Examples of spite in animals may be rare and difficult to distinguish from selfish behavior'. It is difficult to establish exactly why Wilson thought this because he does not explain his reasoning. Although Wilson (1975b) did not directly refer to spite as an adaptation, there is no reason why he would not, when he argued, like Williams (1966, p. 21), that '[a] trait can be said to be adaptive if it is maintained in a population by selection'. Heavily relying on Hamilton (1970), the anecdotal examples that Wilson (1975b, p. 119) does discuss suggests that he thought the ecological circumstances where spite could be adaptive rarely occur, seemingly following the logic of Hamilton's first two points (whilst the omission of any reference to the third point about population extinction suggests that he did not agree with it).

The first example of spite that received more than anecdotal attention was the trait of pirating food and/or attacking neighbouring chicks in the western gull *Larus occidentalis* (Pierotti 1980; Waltz 1981; Gadagkar 1993). There was speculation that piracy could be an alternative foraging tactic for gulls that are (genetically) less successful at feeding (Waltz 1981) or due to harsher environmental conditions (Pierotti 1980). But the main case for spite came from interpreting the piracy behaviour, which was mostly exhibited by males, as costly aggression that risks personal injury to reduce the fitness of neighbours. There is also an alternative case against spite: as males almost exclusively exhibited the behaviour after their offspring for that year had died, the aggression is likely to have low or zero cost for the actor (Keller et al. 1993). Building on this case against spite, the current consensus is that piracy is most likely to be selfish (i.e. not spiteful) in reducing the competition for resources in the actor's reproductive attempts in future years (West and Gardner 2010).

Many of the suggested examples of spite encounter the trade-off of a short-term cost against a longer-term benefit. Perhaps borne out of this frustration, it has been repeatedly suggested that, to enable examples to be more readily found, spite should be defined with respect to the immediate consequences only (most influentially in Trivers 1985, pp. 57–60; but see also Krupp 2013). The difficulty is that it takes inclusive fitness away from a key purpose of using it: to distinguish whether an individual produces a trait for fitness directly through increasing personal reproduction or indirectly through increasing the reproduction of relatives that share genes (West and Gardner 2010). Additionally, many of the general traits that have been suggested to resemble spite using theoretical models such as larger-than-necessary territories, over-the-top aggression, and strategic non-cooperation have incorrectly partitioned the costs and benefits to the actor and recipients (Patel et al. 2020). So there is a clear theoretical argument for clearly defining social behaviours like spite with respect to costs and benefits in the net units of their total fitness effects.

After these first few decades, it seemed well justified that examples of spite would be difficult to find. However, the scope of examples changed with new ideas from the theoretical study of altruism under localised competition (Wilson et al. 1992; Queller 1994; Frank 1998, pp. 114–115). Andy Gardner and Stuart West (2004, p. 1197) presented the theoretical intimations for spite:

> as competition becomes increasingly local, the reference population shrinks towards the size of the social arena, which may contain only a few individuals . . . and/or a significant proportion of identifiable positively related kin, such that the negative relatedness towards the other potential recipients is nontrivial, enhancing the selective value of spite.

Average relatedness remains the same irrespective of the scale of competition (Hamilton 1970; Grafen 1985), so localised competition does not generate negative relatedness in of itself. Instead, if there is an assortment mechanism like genetic recognition, it can give greater rewards for spite because recipients may be genetically very different from the average in the local arena of competition (rather than the whole population). This makes Hamilton's (1970) concerns about the plausibility of spite lessened (if not vanish altogether) because negative relatedness for a rare allele for spite can still have an appreciable value. The general argument in Gardner and West (2004) has stood the test of time (e.g. to be reiterated, albeit with correction to the logic from Hamilton 1975, in Patel et al. 2020).

With an understanding of localised competition, a simple example that is widely repeated to be genuine spite is the soldier caste in polyembryonic parasitoid wasps (Giron et al. 2004; see also West and Gardner 2010). Female wasps lay an egg in a suitable host, such as a caterpillar egg. The wasp is polyembryonic in the sense that the single egg turns into multiple offspring through asexual embryo divisions. These offspring develop into larvae that devour the host from the inside out, emerging as adults to repeat the life cycle. Some parasitoid wasps develop a specialised soldier caste, which have enlarged mandibles for fighting and are incapable of reproduction. As multiple females can lay an egg into a single host, the soldiers serve the adaptive function of defending larvae from the same female and killing larvae from other females using a mechanism of recognition that is currently unknown. The soldier caste is taken as an example of spite because each individual soldier pays the ultimate evolutionary price in being unable to produce offspring, and also because the fighting between soldiers harmfully results in the fatality of their victims.

Whilst the soldier caste might seem to be an unambiguous example of spite because of the hard boundaries of sterility and fatality, there can be more complexity than the forerunning description lets on, which has been explored theoretically in the case of the sterile soldier caste in the parasitoid wasp *Copidosoma floridanum* (Gardner et al. 2007). For this species, female wasps lay one fertilised and one unfertilised egg into the host, and these proliferate into multiple females and males respectively. The females have two developmental paths, one to becoming a reproductive female and the other to becoming a sterile soldier. There are three basic explanations of the existence of sterile soldiers: to defend clonemates, to facilitate host digestion, or to mediate sex ratio conflict. As mating is not restricted to occur before emergence from the host, the last explanation seems the most important. If sterile soldiers were male, acquiring traits that help to defend clonemates or facilitate host digestion would seem more likely because it does not impose a sex ratio cost. Yet, in *C. floridanum*, soldiers are exclusively female, and seem to preferentially kill other males (Giron et al. 2007) to adaptively mediate the sex ratio (as further explained in Hamilton 1967).

Gardner et al. (2007) also includes a lengthy discussion of whether there is any distinction between Hamilton (1970) and Wilson (1975b, p. 119) over the meaning of spite. Kevin Foster et al. (2000, 2001) had previously raised the issue of semantically separating spite from 'indirect altruism', but Laurent Lehmann et al. (2006, p. 1508) had argued this distinction was meaningless in a second major step in the adaptive reformulation of spite (building on Wilson 1975b, p. 119):

> spite against negatively related individuals leads to a fitness increase—and hence altruism—towards the (necessarily positively related) remainder of the population. Inversely, an altruistic trait directed against relatives results in a fitness decrease (spite) in the negatively related remainder of the population. Altruism and spite hence represent two sides of the same coin.

The argument was widely accepted, but as Gardner et al. (2007) remarked (citing Maynard Smith 1976), it comes back to a well-known problem in the philosophy of science about whether it is useful to lump ideas together or split them up. There is a distinction between altruism and spite based on whether or not the recipient is the beneficiary of the interaction, which is true for altruism but not for spite. This is a 'mechanistic' distinction, which is of little relevance if it does not impact how these traits would evolve. Whilst agreeing with Lehmann et al. (2006) that there is nothing to distinguish Hamilton (1970) and Wilson (1975b, p. 119), Gardner et al. (2007, p. 529) goes on to point out that there are important differences about the evolutionary conditions that favour altruism and spite.

A major difference arises in the properties of positive and negative relatedness. This was brought to the fore in Madgwick (2020), where a population genetic approach was used to derive relatedness (that was in keeping with its earlier derivations; see Wright 1922; Hamilton 1964a) whereby the value of negative relatedness from the genetic assortment of individuals with self and nonself alleles is dependent on allele frequency. The analysis assumes global competition, but the results nonetheless suggest that any mechanism of negative assortment would introduce frequency dependence into relatedness. It is the typical pattern that negative relatedness has an inverted U-shaped relationship with relatedness because self and nonself alleles are sufficiently common for many self–nonself combinations to arise (see also Gardner et al. 2004). This differs from positive relatedness, which is typically frequency independent from positive assortment like limited dispersal. Consequently, as David Queller (1994, p. 72) emphasised, 'relatedness is not just a statement about the genetic similarity of two individuals, it is also a statement about who their competitors are'.

It is possible for negative relatedness to arise without frequency dependence from interactions among different positively assorted individuals, such as a trait that harms cousins to help siblings. But then, as Hamilton (1970, p. 1220) also recognised—and had come to prominence in the study of localised competition (West et al. 2002)—the conditions that promote help to one section of the population do not promote harm to other sections. All else being equal, higher relatedness to the third party disfavours altruism but favours spite. Here, the analogy of altruism and spite as 'two sides of the same coin' (Lehmann et al. 2006, p. 1508) breaks down as altruism and spite are not often favoured by different individuals in the same situations, but rather in different situations altogether.

Consider a helper-at-the-nest scenario: imagine an arena of competition where the population is highly mobile so most individuals in the local area are not related. The limited dispersal of offspring from the previous clutch would have high relatedness in terms of an elevated probability of sharing genes with any new siblings that are produced in this year's clutch at the nest. Such limited dispersal would promote altruism in those offspring because the relatedness to their siblings is far from the local average relatedness that is dominated by unrelated competitors. Limited dispersal would not promote spite towards the unrelated competitors outside of the nest because, as Hamilton (1970, p. 1220) put it, 'this leaves a remainder whose average relationship will be only very slightly less (in large populations) than the total average'.

Beyond what Hamilton (1970) discussed, it is also possible to consider a realistic scenario where spite alone may be promoted in social insects with strongly differentiated territories. Territoriality makes the local population in the arena of competition highly viscous, with individuals primarily interacting with related competitors that have a high probability of sharing genes. This biases the local average towards a similarly high probability of sharing genes, and so affords related competitors weakly positive relatedness and unrelated competitors strongly negative relatedness. Whilst indiscriminate spite would be unlikely (see also discussions in West et al. 2002; Patel et al. 2020), targeting spite towards those rarer unrelated competitors would be strongly favoured. Whereas altruism would not be favoured (though is likely to have already evolved beyond easy reversibility for sterile workers in territorial social insects).

An experimental example of the evolutionary relationship between altruism and spite comes from pathogenic bacterium *Pseudomonas aeruginosa*, which can produce iron-scavenging siderophores that have been described as altruistic (Griffin et al. 2004) and conspecific-killing bacteriocins that have been described as spite (Inglis et al. 2011). The siderophore is a straightforward public good that is costly for a bacterium to produce and yet released extracellularly to the benefit of resource acquisition for all in the locality. The bacteriocin is more complicated, being manufactured within a bacterium, but it has to destroy its cell wall, killing itself, to release the toxin into the environment. The bacteriocin gene that produces the toxin also has the capacity to produce an antitoxin, which is expressed by the dying bacterium's clonemates that share this gene to provide immunity from the toxin. As a result, the bacterium's clonemates survive the release of the toxin, whilst conspecific strains are targeted, purging the local area of competitors for resources. Interestingly, experimental evolution under different scenarios of spatial

relatedness and the scale of competition can show that siderophore production is favoured under high frequency (that is like relatedness) and
global competition (Griffin et al. 2004); whereas bacteriocin production is favoured under intermediate frequency irrespective of the scale
of competition (Inglis et al. 2011). If the bacteriocin producer had to
spread from lower frequency, which was not possible in the experimental
design, this was predicted to require stronger local competition. So, with
this one species, it is obvious how different environments in the experiments can favour different traits, but not both traits at once. Further, the
scale of competition can have real consequences for pathogen virulence,
which has also been experimentally demonstrated (Inglis et al. 2009): a
higher diversity of strains within a host can reduce strain frequency to
disincentivise siderophore altruism, leading to greater competition for
resources within the host and to worse health outcomes for the host, or
can give intermediate strain frequency to incentivise bacteriocin spite,
leading to better health outcomes for the host.

With this in mind, it should be no surprise to see that most of the proposed examples of spite are associated with group defence, with altruism
expressed within the group and spite expressed in interactions between
individuals from different groups. Further, it should be of no surprise
that the traits usually involve self-sacrifice so that it is seen as obviously
costly to an actor. Social insects provide many examples (reviewed in
Shorter and Rueppell 2012), but attention must be restricted to those
examples where the group is being defended against members of the same
species, so that the harm that is given is directed at competing alleles
(Frank 1998, pp. 114–115). For example, it is well known that honey
bees (*Apis mellifera*) die when they sting, but such defence is likely to
be evolved against large vertebrate predators, who are more susceptible
to learning the consequences of the penetration of the barbed stinger
(whereas smooth stingers are more effective against other insects). This is
often interpreted as altruism, where the three-party interaction is reduced
to the consequences for the genes in the focal species, so a honey bee
actor is merely helping its colony-mate recipients. By contrast, autoth-
ysis is more commonly directed at raiding members of the same species

in ants, releasing a sticky or toxic substance to eliminate a nest invader. This is spite because all three parties are of the same species. Nonetheless, the reason why both defensive traits persist is much the same, as Wilson (1975b, p. 119) envisaged, with harm to one recipient helping a more related third party.

The more essential job of group defence becomes clearer when considering the toxic rejection trait in the golden star tunicate *Botryllus schlosseri* (Scofield et al. 1982). It is equally true that individual tunicates are susceptible to fragmentation and that juvenile survival is dependent on size, so there are obvious advantages to cooperatively fusing with other individuals in this ecology. However, if individuals are unrelated, fusion is aborted in favour of a toxic rejection that usually results in the death of one or other of the interactants. Here, the idea of group defence is taken to its logical conclusion as a kind of immune system that maintains a cooperative group against external threats.

The immune system is widely regarded as one of the most impressively complex adaptations, and so all of these reputable examples of spite naturally sit within an adaptive interpretation (Madgwick et al. 2019). With this analogy, it would be difficult to argue that, for group defence behaviours, 'the selection of a gene causing spite can only hasten [a population's] extinction' (Hamilton 1970, p. 1220). Instead, by the interpretation of the examples that have been accepted, Wilson (1975b, p. 119) has implicitly won the argument that spite is understood as providing a benefit to a third party. But there is some merit in distinguishing the ideas of Hamilton (1970) and Wilson (1975b, p. 119). Hamilton (1970) makes no mention of third parties (or secondary recipients). Given his inference that spite should hasten extinction, it is reasonable that Hamilton had a concept that was closer to maladaptation than the current meaning of spite, which is closer to the original conception of Wilson (1975b, p. 119).

Accepting this distinction, on the basis of evidence, there is still room for doubt that these best examples of spite are necessarily adaptive. These traits share an obvious cost to an actor in their death and an obviously harmful effect on a recipient to define them as spite, but the only inference of the effect on individual fitness comes from there being positive inclusive fitness, which must be the case for the trait to be favoured by natural selection. But whether a trait is adaptive or maladaptive is a completely separate issue to the concerns of inclusive fitness; it is conceivable that any the form of social behaviour, which is typically classified as altruism, spite, selfish, or mutual benefit (West et al. 2007a; see also Marshall 2015), could be a maladaptation when the costs and benefits that define inclusive fitness describe the net effect on the relative replication of the gene that causes a social behaviour. Adaptation and maladaptation are defined based on absolute fitness, but inclusive fitness is built on relative fitness.

Outside of theory, inclusive fitness is sometimes used with an implicitly alternative definition based on absolute fitness, especially in experimental studies that rely on fitness proxies. The studies of spite have so far been incapable of approximating the costs and benefits, but a classic example of altruism can be provided from helping at the nest in the Florida scrub jay (*Aphelocoma coerulescens*) (Mumme 1992; see also Ridley 1993, pp. 299–301) to demonstrate this kind of quantification. Adult offspring from the previous year may sometimes stay at the nest to help rear their parents' juvenile offspring in the following year. Ronald Mumme (1992) found that the presence of two helpers at the nest increased juvenile offspring survival from 7% to 35%, so a single helper can be attributed a $+14\%$ increase in juvenile survival. This describes the benefit to the recipient, who is a sibling that shares 50% of the helper's genes. Costs are harder to calculate but can be estimated within bounds. If an adult offspring would have been unable to breed on their own then the cost would be zero; if they could obtain a mate and a breeding site then, like their parents, the survival of their own juvenile offspring in the absence of a helper would be 7%. Their own offspring also share 50% of their genes, so the marginal inclusive fitness of one trait or another can be set up by contrasting the 14% increase in survival from helping their siblings to reproduce against what is at maximum only a 7% increase in survival from attempting to produce their own offspring; hence helping at the nest is favoured. This proxy of fitness is in absolute units of juvenile survival, which is an experimental measurement that may be used to detect maladaptation in other examples.

Whilst there is hope that experimental studies have used a proxy of inclusive fitness that could differentiate adaptation and maladaptation, a key problem in using the literature on spite to find examples of maladaptation is that most examples are anecdotal. It is typical to have an obvious harm to the actor due to personal risk or self-sacrifice, and an obvious harm to the recipient due to some resulting injury. But these harms must be translated into the total fitness costs and benefits of inclusive fitness, which may for example introduce a probability of self-sacrifice that makes its costliness less binary. Moreover, whilst empirical studies have often measured absolute inclusive fitness, the use of fitness proxies cannot rule out alternative pathways to fitness, for example in the longer run. With considerations like these, it is hard to find uncontentious evidence of maladaptation among examples of spite.

Yet, many of the examples of traits that have been proposed and dismissed as examples of spite (see West and Gardner 2010) exhibit a harmfulness that could express maladaptation. Whilst they may not provide a reliable set of examples, such traits may well merit further investigation. For example, three-spine sticklebacks *Gasterosteus aculeatus* coexist with a related species of black-spotted sticklebacks *G. wheatlandi*, which are much more common (Gadagkar 1993). Whilst very similar species, female three-spine sticklebacks differ in attacking the nests of members of their own species to eat their eggs. In experimental study, it could be shown that female three-spine sticklebacks prefer to eat conspecific eggs over black-spotted stickleback eggs, which is surprising when they are nutritionally identical. It could be that the eggs were produced by females that have a similar diet, leading to the expression of a food preference (Keller et al. 1993), but this would not explain why three-spine sticklebacks do not attack black-spotted stickleback nests. Whilst it may be that conspecific egg cannibalism is selfish because it reduces the subsequent competition for a female's own offspring (West and Gardner 2010), the basic difference in the behaviour and population sizes of the two related sticklebacks suggests that this example may still be a case of maladaptation. Additional study is required to estimate the absolute fitness effects to take this example beyond being merely suggestive of maladaptation.

* * *

The third pitfall is taking genetic conflict to imply maladaptation. Genetic (or intragenomic) conflict has received numerous definitions (reviewed in Gardner and Úbeda 2017). The concept of conflict emerged from the discussion of traits in groups of individuals, where it is useful to distinguish potential and actual conflict (Ratnieks and Reeve 1992): potential conflict arises with a disagreement over a trait optimum, whereas actual conflict arises when an individual dose something about such a disagreement. The earliest general definition of genetic conflict applied the notion of potential conflict to genes in the same genome: 'There is a genetic conflict if the spread of one gene creates the context for the spread of another gene, expressed in the same individual, and having opposite effect' (Hurst et al. 1996). Gardner and Úbeda (2017) have argued that this definition is too permissive in also applying to adaptive fine-tuning, where genes may not disagree about the optimum but produce phenotypic effects in different directions to get closer to an optimum. Arguably, this misinterprets 'having opposite effect' in reference to a continuous phenotype rather than with respect to an optimum, but Gardner and Úbeda (2017) nonetheless work towards the same conclusion by rephrasing the definition in their own terms. Critically, a genetic conflict is defined by the potential for an allele at one locus to change the phenotypic effects of an allele at another locus to increase its own replication at the expense of the other allele.

A wide diversity of actual genetic conflicts have been discovered, including those that play out in the context of social behaviours like genomic imprinting (as reviewed in Burt and Trivers 2006). Over the years, one of the most active areas of discussion has been over so-called greenbeard genes. The original concept stemmed from Hamilton (1964a, p. 25):

> That genes could cause the perception of the presence of like genes in other individuals [which] may sound improbable; at simplest we need to postulate something like a supergene affecting (a) some perceptible feature of the organism, (b) the perception of that feature, and (c) the social response consequent upon what was perceived.

Dawkins (1976, p.89) described the evolution of such a signal–receptor–behaviour gene as via a 'greenbeard effect': 'It is theoretically possible that a gene could arise which conferred an externally visible "label", say a pale skin, or a green beard, or anything conspicuous, and also a tendency to be specially nice to bearers of that conspicuous label'. Both Hamilton and Dawkins agreed that this was highly improbable because it required these three particular traits to come together in one gene:

> Green beardness is just as likely to be linked to a tendency to develop ingrowing toenails or any other trait, and a fondness for green beards is just as likely to go together with an inability to smell freesias. It is not very probably that one and the same gene would produce both the right label and the right sort of altruism. (Dawkins 1976, p. 89.)

Nonetheless both thought greenbeards were a useful thought experiment for explaining the logic of altruism.

In particular, the greenbeard logic helped to explain why altruism between individuals is selfish *for a gene*. Dawkins (1976, p. 89) used a greenbeard as his first description of a gene that was not only potentially selfish because it was favoured by natural selection to increase its own rate of replication, but actually selfish because it produced a trait that could do something about it. The logic was meant as a thought experiment to crystallise the idea of a gene in one individual helping itself to replicate by being altruistic to another individual that it can be sure also carries the same gene. Dawkins (1976, pp. 90–91) then swiftly moves on to discuss the more complex case of altruism among relatives, which have a greater than average chance of sharing genes through common ancestry. Going onwards, this is particularly framed in the most plausible context without the genetic recognition of a greenbeard, instead relying on inherent associations built up by behaviours within a family context or when kin have limited dispersal.

Over the years, the arguments against the existence of greenbeards have mounted (as reviewed in Madgwick et al. 2019). Besides it being unlikely that just the right pleiotropy would come together to bring about the three traits, the argument against the existence of greenbeards has been driven further, as it seemed unlikely that a gene could deterministically produce the presumably complex signal traits, when development often introduces environmentally-influenced variations into traits like colourations (Dawkins 1982, pp. 143–146; Zhang and Chen 2016). Selection arguments also arose that even if a greenbeard could exist, they would be disfavoured by selection because either the advantage of a cost-free greenbeard signal could be eliminated by the presence of a cheater that displays the signal but does not produce the behaviour (Gardner and West 2010; Biernaskie et al. 2011), or a greenbeard with a costly signal might be unable to invade when it scarcely encounters itself to reap the benefit of the costly signal when it is rare (Biernaskie et al. 2011, 2013). Lastly, detection arguments arose that even if greenbeards could exist and were selected, they would be impossible to detect because either they would rapidly spread to fixation (and so cease to be displayed; Gardner and West 2010), rapidly be silenced by other genes in the genome (Alexander and Borgia 1978; Ridley and Grafen 1981), or they would drive the host species to extinction (Hamilton 2001, pp. 331–326). This last argument relates to genetic conflict and maladaptation. A greenbeard reaps its benefit because the signal is associated with the presence of the greenbeard gene, but that same signal may have no associations with the presence of other genes. Consequently, whilst every gene in the genome pays the costs of the greenbeard traits, only the greenbeard reaps the benefits, generating the potential for genetic conflict.

Despite all these arguments against the reality of greenbeard genes, it is perhaps surprising that examples have now been proposed (as reviewed in Madgwick et al. 2019). This has led to a reevaluation of the existential arguments against them. When greenbeards were first proposed, the relationship between genes and traits was less clear, especially in animals, which were the main focus of discussion at the time. Many of the suggested examples of greenbeard genes have been enabled by the advances in molecular biology, coming from cell-surface receptors (Haig 1996) or linked signal–receptor pairs (Madgwick et al. 2019) that are uniquely placed to govern the signalling and response by the movement of molecules through cell membranes to govern a social behaviour. Although many of the best examples also come from microbial organisms (Gardner and West 2010), the key enabler of the discovery of greenbeards has been a reconsideration of the scale of interaction rather being tied to the increasing interest in microbes (Madgwick et al. 2019). For instance, there are also potential examples of greenbeards from complex organisms including humans for social behaviours governed by molecular interactions (e.g. cadherins that govern maternal–foetal interactions; Haig 1996).

The selection arguments also made assumptions that do not hold in this new wave of examples. The selection arguments assumed that greenbeards operated in well-mixed populations, so that the only way that two individuals could share genes is through the greenbeard mechanism (Madgwick et al. 2019). Every newly proposed example of a greenbeard seems to naturally take place in a structured population, so there is a high probability that individuals who share the greenbeard also share other genes throughout the rest of the genome by kinship. Such background relatedness, which has a similar logic to the ecological effect (see Chapter 3), can prevent cheaters and enable costly signalling by increasing the frequency with which individuals with different alleles interact—and so benefit from the greenbeard. This works through much the same logic as other traits that evolve under kin selection (Grafen 1985). Arguably, in as far as this applies to each example, kinship remains the key driver behind the evolution of social behaviour in nature (see West et al. 2007b). But even if kinship adds some incentive to the other genes in the rest of the genome, greenbeards still reap a disproportionate benefit by guaranteeing the recipient shares the greenbeard allele rather than sharing any allele according to the degree of kinship (Madgwick et al. 2019).

The detection arguments against greenbeards remain broadly valid, but have been skilfully circumvented by experimental ingenuity. Greenbeards have been easily detected in experiments that bring together individuals from subpopulations that may rarely interact in nature in order to identify the greenbeard effect (e.g. Gruenheit et al. 2017). Consequently, even if a greenbeard is fixed within a subpopulation, it is detectable in the laboratory (Queller et al. 2003). Similar to the selection arguments, background relatedness may also play an important role in explaining why genes in the rest of the genome do not benefit from silencing the greenbeard effect (Madgwick et al. 2019); indeed, no such modifiers have yet been discovered for any of the examples, presumably because they benefit from the greenbeard's effect through kin selection.

The error in the last of the detection arguments has been most puzzling: why do greenbeards not drive their host species to extinction? Almost all commentators on greenbeards have seemed totally unaware of Hamilton's expectation that, when he proposed the underlying concept, he expected greenbeards to drive host species extinction (as clarified in Hamilton 2001, p. 326). Hamilton (2001, p. 326) states that he had never imagined a real example being found because it 'is rather as one half of the theory predicts; but then why isn't the other half of the theory working, the half that says that the green-beard gene complex should be killing its host species off and for this reason isn't found?' (see also Hamilton 1987). He was insistent that the other half of the theory was not wrong—and should be working (Hamilton 2001, pp. 331–332) as 'a genome that is perfectly "fair" to all kinds of fitness advantage, whether disruptive or not, couldn't be anything like what we would today call an organism—it could be a leaderless gang of DNA bits and pieces, a minute, weak-membraned proto-cell of the primordial ooze perhaps, nothing more', such that 'by the green-beard gene's increase to 100% . . . our focal species is greatly weakened and very likely to become extinct'. Indeed, Dawkins (1982, pp. 250–264) seems equally puzzled after 'rediscovering the organism' and identifying the 'paradox of the organism' (Dawkins 1990, p. 64): 'Why is the organism not torn apart by the conflicting interests of the multitude of self-interested units that it contains?'. Dawkins (1990, p. 71) explains his general solution with reference to parasites that transmit from parent to offspring: 'The important fact about an organism's own nuclear genes is that they all share the same gametes. It is for this reason, and this reason alone, that they stand to gain from the same set of outcomes in the future'. But alas this does not help explain the case of the greenbeards that have vertical transmission.

To take an example, the first proposal of a greenbeard gene was at the *gp9*-locus in the fire ant *Solenopsis invicta* (Keller and Ross 1998). It is typical in ant species that virgin queens leave the nest to find a mate and disperse to find a suitable location to found a new colony. In *S. invicta*, mated queens often return to the nest they came from to form a polygynous colony with multiple mated queens. When they return to their home nest, workers can either take the queen into the colony or kill them. In a genetic study of the accept-or-reject behaviour, the probability of attack was heavily dependent on the genes of the individuals involved. In the laboratory, workers that were heterozygous at the *gp9*-locus were largely responsible for attacks, and overwhelmingly targeted homozygous queens that did not possess a particular allele in the workers. To check for a genetic cause, this could be further tested by contrasting newly mated and older queens, where the same pattern was observed. Additionally, as a further check, cuticular chemicals could be transferred to queens, and the probability of an attack could be manipulated. The identified allele has been interpreted as a greenbeard gene for altruism towards accepted queens (Grafen 1998), but some argued that it should be viewed as spite against the rejected queens (Hurst and McVean 1998). Both phenomena appear to be inseparable features of the same trait.

There are several features of this example that may explain its occurrence with respect to the existential, selection, and detection arguments. The existence of the greenbeard owes to a chromosomal inversion, which has created a large non-recombining region containing many linked genes (Wang et al. 2013; as seemingly predicted by Dawkins 1982, p. 149). The inversion contains protein-coding regions for the production of up to nine odour-binding molecules that are expressed in the cuticular hydro-carbon profile of the workers and queens (Pracana et al. 2017). Such a molecular greenbeard signal is likely to be cheap to produce (in comparison to an actual green beard). Further, the accept-or-reject behaviour takes place in the context of queens returning to the nest that they came from, and so returning queens are likely to share some fraction of their genes with the workers that decide the course of action (Keller and Ross 1998); the background relatedness may prevent the spread of cheating the signal. Behind the detection of this example, it is important to note that homozygotes with the greenbeard allele appear developmentally incompetent, suggesting that the greenbeard is recessively lethal (Keller and Ross 1998). Consequently, it could not spread to fixation within the population. Moreover, an analysis of native populations of *S. invicta* has revealed that there are multiple greenbeard alleles that control the same behaviour with a different signal and receptor (Trible and Ross 2015), which is like having multiple beard colours. Many of these features are shared with the other proposed examples of greenbeards (Madgwick et al. 2019).

But there is still that nagging question: why does the greenbeard persist without driving the population to extinction? Hamilton (2001, pp. 331–336), discussing this example, is insistent that greenbeards are like parasites that can transmit themselves through a population whilst introducing a disease-like cost on its host individuals. Instead of doubting his expectation, Hamilton (2001, p. 326) finds an alternative answer to why the greenbeard not killing off its host species, relying on the idea that

> an adaptation that is definitely selfish and destructive at one level of grouping of genes may provide a benefit at a higher group level. The modest numbers of 'green-beard' executions of queens that occur due to the gene may facilitate a type of organization that is proving very beneficial at the colony level and above.

Hamilton (2001, p. 326) goes on to suggest that greenbeards may be more common than he initially supposed if—and seemingly only if—they are 'beneficial to the overall density of population' (cf. Williams 1966, pp. 26–28).

Hamilton's (2001, p. 336) suggestion is overly simplistic because very few details were known about this example at the time, but it is nonetheless revealing about what theoretical oversight he saw in his original expectations about greenbeards. Hamilton, like others who viewed greenbeards as outlaws (e.g. Alexander and Borgia 1978), had assumed that greenbeards were damaging to an individual's fitness because there was the potential for genetic conflict between the greenbeard and genes in the rest of the genome. But the potential for genetic conflict is not the same as a decrease to absolute fitness. Indeed, it has been argued that a greenbeard only spreads because it increases an individual's relative fitness (Ridley and Grafen 1981), albeit that this reasoning ignores the potential for a cheater allele to increase its relative fitness at the expense of a greenbeard allele (Gardner and West 2010). Interestingly, no examples of cheater alleles have been found for the most convincing examples of greenbeards that have been proposed to date, which is likely to be because of interactions among kin with background relatedness (Madgwick et al. 2019). This all adds weight to Hamilton's (2001, p. 336) correction of his previous reasoning, which suggests that identifiable greenbeards may generate a net increase in absolute fitness, like any other adaptation. With similarities to many of the examples of spite (that rely on a form of genetic recognition like greenbeards), the most convincing greenbeards appear to have a functional role in group defence against members of the same species. Indeed, currently, there are no examples that do not fit this description.

Arguably, the best example of a greenbeard to date comes from the slime mould *Dictyostelium discoideum* in the steps of its multicellular fruiting body development, which has already been discussed in passing. As free-living cells aggregate together under starvation conditions, groups of cells may segregate away from one another or stay together to undergo fruiting-body development, which involves some cells altruistically becoming non-reproductive stalk cells to enable the other cells to disperse as spores (Strassmann et al. 2000). The segregation behaviour is controlled by a polymorphic cell-surface receptor at the *tgrB1/tgrC1* locus that causes cells to stick to one another, with TgrB1 acting as a receptor for the TgrC1 ligand (Benabentos et al. 2009). It is a greenbeard rather than kin recognition because the only predictor of cells sticking together is matching at the *tgrB1/tgrC1* locus (Gruenheit et al. 2017). But, of course, the population structure after many rounds of clonal reproduction gives cells high background relatedness (Buttery et al. 2012); unsurprisingly, no cheater *tgr* alleles have ever been found. The greenbeard behaviour is much more sophisticated than cell adhesion, as it also controls the allocation of cells into the stalk or spores (Madgwick et al. 2018). As strains of cells within an aggregation adjust their cell allocation, one strain may cheat another; a rarer strain can benefit from a lower stalk cell allocation without inflicting too much of a cost on forming the stalk. When both strains are at intermediate frequencies, they may try to cheat each other, which results in the catastrophic failure of stalk production (Belcher et al. 2022). In this way, the segregation behaviour that is controlled by the *tgrB1/tgrC1* locus is like an immune adaptation, in preserving the multicellular fruiting body against social parasites in the strains that would attempt to be cheaters by biasing the allocation of their cells towards the reproductive spores (and away from the non-reproductive stalk cells). In this way, it seems likely that the full suite of segregation and allocation behaviours that are controlled by the *tgrB1/tgrC1* greenbeard give a net increase to a cell's absolute fitness.

Therefore, the potential for genetic conflict does not imply maladaptation, but instead describes a separate phenomenon. Genetic conflict arises from the potential for an allele to increase its own replication at the expense of one or more other alleles in the genome. Maladaptation may occur because of genetic conflict, but only when the potential is actualised in a way that harms the absolute fitness of individual organisms. Genetic conflict can lead to other outcomes including adaptation, which seems to be the case for the greenbeards that have been identified to date.

Overall, the three pitfalls have a common theme in the mistaken supposition that an identified kind of harmfulness is relevant to the absolute fitness that defines maladaptation. The first pitfall comes from assuming that intraspecific competition is bad for everyone, whereas it can beneficially allocate resources to those that are best placed to use them. The second pitfall comes from misunderstanding the relationship between inclusive fitness and individual fitness, where it becomes apparent that spite and other Hamiltonian social behaviours are defined based on their relative effects rather than absolute fitness. The third pitfall comes from mistakenly thinking that genetic conflict implies that genes like greenbeards are maladaptive, when the identification of a genetic conflict—as a potential—is not evidence that it actually harms an individual's absolute fitness. More often, identified greenbeards seem to involve adaptation in group defence. Overall, then, perhaps it is surprising that when reviewing candidate social behaviours, there are not many examples that convincingly appear to be maladaptations. A critical issue is that, because it was not of interest to them, studies have rarely collected persuasive demonstrations of decreasing absolute fitness—especially under natural conditions. Building on the discussion of genetic conflict, the next chapter moves on to consider maladaptation within the body, where these details tend to be more obvious.

5

Maladaptation within the body

Maladaptation within the body is currently where the most persuasive examples of maladaptation are found. The body is often thought of as the exclusive domain of a single individual, but it is a social environment for the genes that make up that individual. Indeed, in many of the most familiar organisms, a key part of the social nature of the body stems from diploidy, where each individual inherits two copies of each gene, one from their mother and one from their father. Those genes need not be the same, representing different alleles that are competing to replicate into the same locus in their offspring, as each parent only passes one of their alleles on to each offspring. Consequently, there can be intense competition between the maternal and paternal alleles to secure their place in the next generation. With such high stakes, the alleles can afford to resort to extreme tactics that can be extraordinarily costly to an individual's fitness in the number of offspring that they produce. But whilst particular offspring may suffer even unto the point of death, the total harm to all offspring can often be mitigated through reproductive compensation in reallocating resources from the production of some offspring towards others, which generates an ecological effect (analogous to density dependence among social behaviours between individuals) that makes maladaptation much more likely to evolve. In this way, the diploid organism is an almost perfect environment for the localised competition that permits maladaptation, allowing slim advantages to be gained despite high levels of collateral damage to organism design.

A classic example of allelic competition within the body—and what is expected to be considered as one of the clearest examples of maladaptation—comes from meiotic drive, which involves the non-Mendelian transmission of genes between the generations. Meiotic drive has been known about for a long time (Gershenson 1928), including the possibility that it would enable the spread of disadvantageous traits (Haldane 1932, p. 123) that may even drive species to extinction (Sandler and Novitski 1957). An allele for meiotic drive can be thought of as much like greenbeard, already discussed (see Hamilton 1964b), using a signal that it can detect to control an effect. A canonical mechanism is an autosomal killer via a toxin–antitoxin mechanism acting on sperm cells (Burt and Trivers 2006, p. 3). The drive allele produces a toxin that kills sperm cells that do not have the drive allele, whilst sperm cells that do have the drive allele produce an antitoxin that neutralises the toxin's effects. The result is that the individual tends to produce sperm that pass on the drive allele into their offspring. The mechanism is more commonly expressed in males than females because reproductive compensation—that takes on the role of the ecological effect within the body—tends to be cheaper, as males tend to contribute fewer resources to offspring production. Nonetheless, meiotic drive can lead to infertility through numerous mechanisms, including through the direct action of the toxin or other indirect associations with linked alleles (Zanders and Unckless 2019). Various different examples have been found across the tree of life in animals, plants, and fungi (as reviewed in Burt and Trivers 2006). Indeed, there are even examples in humans, albeit that these examples are relatively poorly characterised owing to the obstacles of the ethical study of such phenomena.

One of the best-studied examples of meiotic drive comes from the tail (*t*) haplotype in the house mouse *Mus musuculus*. The *t* haplotype was originally identified because it causes a small tail in adult mice, which is an obvious trait in laboratory populations (Chesley and Dunn 1936). Since its identification, numerous allelic variants of the *t* haplotype have been discovered in natural populations; for example, 16 variants were identified in different European populations (Klein et al. 1984). The vast majority of variants tend to be recessive lethal, meaning that an individual cannot survive carrying two copies of the *t* haplotype. In heterozygotes, around 90% of fertilisations pass on the *t* haplotype from a father to their offspring, but this can be lower in some populations, due to some resistance to its effects (Gummere et al. 1986). Subpopulations show a wide range of frequencies of the *t* haplotype between zero and 71%, but the total frequency in different populations is usually close to 5% (Ardlie and Silver 1998). Resistant alleles tend to be much rarer in natural populations (Ardlie and Silver 1996). For a long time, the mechanism of its operation was unknown, but the molecular details have been gradually uncovered. The *t* haplotype is a large non-recombining region that accounts for close to 1.2% of the mouse genome, which has persisted for over 2 million years in the mouse lineage (Silver 1993). The *t* haplotype achieves meiotic drive by disrupting the Rho GTPase signalling cascade to impede the motility of sperm that do not carry the *t* haplotype (Bauer et al. 2005). Experimental study that mimicked the natural ecology estimated that the fitness reduction to heterozygotes is likely to be around 21% for females and 36% for males, or greater if there is inbreeding (Carroll et al. 2004). Whilst the drive mechanism is undoubtedly harmful in reducing the number of competent sperm with or without the *t* haplotype (Herrmann et al. 1999) directly resulting in lower fertility (Sutter and Lindholm 2015), the large non-recombining region also contains strongly deleterious alleles that are inessential to the drive mechanism (Schimenti et al. 2005). With such strong links to infertility, the *t* haplotype is a clear example of maladaptation.

Other examples of meiotic drive can have different evolutionary properties. Some meiotic drivers exhibit rapid evolutionary turnover in arising by mutation, spreading to high frequencies, and then declining to be lost from the population, whilst others persist for a very long time (Price et al. 2019). An interesting example is the with-Tf-retrotransposons locus (*wtf*) in the fission yeast *Schizosaccharomyces pombe*, so named because it is found near the long terminal repeats derived from Tf retrotransposons (Bowen et al. 2003). Instead of there being just one driving element, there are as many as 32 identified *wtf* genes in different strains of fission yeast, with some strains having more than ten active *wtf* alleles and many inactive pseudogenes that are no longer driving (Hu et al. 2017). An active *wtf* allele can bias its transmission into the next generation at a rate in excess of 70% through a poison–antidote mechanism (Nuckolls et al. 2017). When multiple *wtf* genes are mismatched in crosses between distantly related strains, multiple *wtf* alleles have been implicated in the resulting complete infertility (López Hernández and Zanders 2018). Whilst many of the pseudogenes appear to have been much shorter-lived, the *wtf* family as a whole has been estimated to have persisted for over 100 million years in fission yeast (De Carvalho et al. 2022). Despite such different evolutionary dynamics, *wtf* is another example of maladaptation.

Meiotic drivers generate strong selection that can lead to their rapid fixation, which can make it difficult to detect their presence based on the drive phenotype. The hidden presence of meiotic drivers is often revealed during hybridisation (Hurst and Werren 2001). For example, the cultivated rices *Oryza sativa* and *glaberrima* may both exhibit the Mendelian inheritance of traits in crosses within the species, but there is a pollen killer (called *S2*) that is fixed (i.e. at 100% frequency) in *O. sativa* that causes hybrid sterility with *O. glaberrima*. The surprising frequency at which meiotic drivers are found in hybridisation tests that can discern them has led to many statements of the importance of meiotic drivers to the evolution of organism design (Sandler and Novitski 1957; Zimmering et al. 1970; Werren et al. 1988; Hurst et al. 1996; Hurst and Werren 2001; Rice 2013; Ågren and Clark 2018). Indeed, the evolution of 'intrinsic postzygotic isolation' from elements like meiotic drivers has recently been recognised as an important and maladaptive process, in the sense of natural selection favouring the production of unfit offspring (as reviewed at length in Coyne and Orr 2004, pp. 247–319). In this way, meiotic drivers play a part in a maladaptive theory of speciation (that contrasts against an adaptive theory based on local specialisation or a non-adaptive theory based on extrinsic isolation). Here, the harm caused by the meiotic driver extends from whatever costs there may be from carrying a driver without transmission distortion, to the resulting reduction of the pool of compatible mates that comes from being unable to produce fit offspring when there are intrinsic barriers.

Meiotic drivers are a good example of the kind of evidence there is for maladaptation within the body. Although there is no obvious way that meiotic drivers are good for the organism, their effects on the whole organism can be surprisingly complicated and difficult to quantify. Yet, through the painstaking experiments that have established the molecular details of how meiotic drivers work and have evolved, it appears that they have been long-lived phenomena that harm the fertility of organisms. As such, meiotic drivers often provide clear examples of maladaptation.

* * *

Beyond meiotic drivers, there are many other examples of what have become known as 'selfish genetic elements' (Werren et al. 1988) that enhance their own transmission into the next generation at the expense of other genes in the rest of the genome. The wide diversity of mechanisms that these elements use to replicate outside of Mendelian inheritance have been influentially reviewed in *Genes in Conflict* by Burt and Trivers (2006), including many mechanisms that will be overlooked here such as genomic imprinting, gene conversion, and B chromosomes. The review finishes with a strong conclusion about their assessment of the nature of adaptation in the body (Burt and Trivers 2006, p. 475, emphasis original):

> The unity of the organism is an approximation, undermined by these continuously emerging selfish genetic elements with their alternative, narrowly self-benefiting means for boosting transmission to the next generation. The result: a parallel universe of (often intense) socio-genetic interactions *within* the organism—a world that evolves according to its own rules, as modulated by the sexual and social lives of the hosts and the Mendelian system that acts in part to suppress them.

Whilst this sounds promisingly supportive of the concept of maladaptation that is presented here, there is much to unpack within the conclusion, whereupon it becomes apparent that Burt and Trivers are much more sceptical of selfish genetic elements causing maladaptation than they might initially seem.

Although it is little remarked upon, one evident finding from *Genes in Conflict* is that the quantification of the harm that selfish genetic elements inflict on organisms is often poor. Indeed, others have made the case that some selfish genetic elements do not harm their hosts at all (e.g. selfish centromeres; Henikoff et al. 2001). Regardless of the quantity of harm, Burt and Trivers (2006, p. 471) maintained that it is logical that 'The immediate, short-term effects of selfish genetic elements on their host organisms are negative', though their 'effects on mean fitness are not to be confused with effects on population productivity'. In this way, Burt and Trivers seem to recognise a distinction between the relative fitness that leads to a gene being favoured by natural selection and the mean (i.e. absolute) fitness that leads to that gene becoming more numerous in the population. Moreover, Burt and Trivers (2006, p. 448) also suggested that the long-term effects of selfish genetic elements may even be beneficial to their host organisms. For example, they speculate that a driving X chromosome may lead to an increase in population size by producing more females that, assuming that the fewer males can mate with all the females, have more offspring per generation (Burt and Trivers 2006, pp. 11–12). By implication, perhaps unequal sex ratios could be adaptive consequences of selfish genetic elements.

Nonetheless, more commonly, selfish genetic elements are presented as capable of reducing mean fitness to drive a population towards extinction (Burt and Trivers 2006, p. 471). For opposite reasons to a driving X chromosome, a driving Y chromosome may lead to a decrease in population size by producing more males. Natural examples of driving Y chromosomes have been discovered in two species of mosquito, *Aedes aegypti* and *Culex quinquefasciatus*, leading to an excess of males in the population (Wood and Newton 1991). Further, a driving Y chromosome has been shown to rapidly cause extinction in the laboratory populations of *Drosophila melanogaster* (Lyttle 1977). Furthermore, theoretical modelling has shown that extinction is more likely for driving Y chromosomes than driving X chromosomes (Hamilton 1967), which may help explain why driving Y chromosomes are so much less commonly found in nature than driving X chromosomes (Burt and Trivers 2006, pp. 73–74). The speed of a driving Y chromosome causing population extinction may make a natural example difficult to detect. An example where extinction would seem likely to occur (without human intervention) comes from androgenesis, which has been implicated in the decline of the tarout tree *Cupressus dupreziana* (Burt and Trivers 2006, pp. 416–418). At present, only 10% of the seeds contain a viable embryo because of a selfish genetic element (Pichot et al. 1998). The population may be as low as 231 individuals, with the majority of trees being estimated as over 2,000 years old (Pichot et al. 2001).

With such overlapping support, there seems to be scope for interpreting maladaptation in organisms as a possible consequence of selfish genetic elements within them. But there are aspects of the ideas in Burt and Trivers (2006) that run counter to this interpretation. It is important to recognise the language that is being used. Throughout *Genes in Conflict*, Burt and Trivers prefer to think of organisms as hosts and selfish genetic elements as their parasites, which is in keeping with the tradition from Dawkins (1976). The host–parasite language seems most natural for those selfish genetic elements that can replicate themselves separately from the rest of the Mendelian genome, like a transposon with horizontal transmission to create a new copy of itself within the genome. In these cases, there is clearly an entity that is replicating like a parasitic species. For the majority of examples, selfish genetic elements replicate alongside a full set of other genes in the Mendelian genome, which is very unlike a parasite. Interestingly, there are examples of parasites that do replicate through vertical transmission from parents to offspring like the intracellular bacteria *Wolbachia*, but these are exceptional. By choosing to use the host–parasite language indiscriminately for all organisms and their selfish genetic elements, there appears to be a tacit acceptance that the difference between a host and their parasite comes down to way that a parasite may benefit from harming their host. This distinction becomes trickier to maintain with the new concept of maladaptation because there is no reason to think that what is 'of the organism' is always 'for the organism'.

The host–parasite language becomes important to the theory of how selfish genetic elements persist. Burt and Trivers (2006, p. 457) discuss selfish genetic elements as 'the original sexually transmitted diseases'. The identified importance of sex to the spread of selfish genetic elements is hardly unique to this review (e.g. Hurst and Werren 2001), but it is interesting that the argument is maintained alongside the way that *Genes in Conflict* recognises a much wider diversity of selfish genetic elements than any review before it. Indeed, the introduction and summary of the book (Burt and Trivers 2006, pp. 10–11 and 457–458) include statements about how selfish genetic elements may persist by spreading between species during hybridisation events, but this seems to be of very limited importance to explaining the persistence of the vast majority of the examples that are discussed. Moreover, Burt and Trivers (2006, p. 458) discuss 'an inability of selfish elements to spread through an asexual population' because 'the entire genome is transmitted intact from one generation to the next and acts as the unit of selection'; consequently, 'different [clonal] lineages will inevitably be differentially burdened and there will be selection among lineages for those that are less burdened'. This description ignores how bacteria have toxin–antitoxin cassettes that operate in very similar ways to many meiotic drivers—and can spread through a population by horizontal transmission from individual to individual independent of the genome of origin (see Niehus et al. 2021 for a review). Here, it seems, the host–parasite language restricts the understanding of the theory of the relationship between selfish genetic elements and sexual reproduction in a way that limits its scope.

The peculiarities of the host–parasite language and their view of the relationship to sex lead on into the crux of the problem with associating selfish genetic elements with the concept of maladaptation; Burt and Trivers (2006, p. 464) argue that 'Whether a selfish genetic element goes to fixation in a population or remains at some intermediate equilibrium frequency, over the long term it will almost certainly go extinct within that species. If such extinctions did not occur, all species would have many more selfish genetic elements than they do'. Three arguments are presented in support of this conclusion, but only one is particularly persuasive. First, in specific terms, 'An autosomal killer that goes to fixation has no susceptible alleles left to attack, and so there is no longer selection for killing' (Burt and Trivers 2006, p. 464). An absence of selection appears assumed to imply the loss of the selfish genetic element, but there may be no way for the selfish allele to degenerate to restore a non-selfish allele because the selfish allele would always outcompete any non-selfish competitor (Madgwick and Wolf 2021). Further, even though drive may not occur, the selfish genetic element may still impose fitness costs on fertility (Zanders and Unckless 2019). Second, specifically for large non-recombining regions, there can be 'the accumulation of deleterious mutations and the failure to incorporate new beneficial mutations' (Burt and Trivers 2006, p. 465), which could lead to selection against the selfish genetic element. Of course, this could happen, but equally the non-recombining region could accumulate coadapted mutations, which may partially explain the persistence of other large non-recombining regions like sex chromosomes over very long timespans (Abbott et al. 2017). Third, most persuasively, Burt and Trivers (2006, p. 465) argue that selfish genetic elements may 'accumulate their own parasites—defective versions of the element—that swamp out the functional ones', leading to 'host genes that are resistant to or suppress the activity of the selfish gene'. This is the most convincing of the three arguments because, for many of the examples in *Genes in Conflict*, some general or specific mechanisms of suppression have been identified.

Consequently, despite establishing their wide diversity, Burt and Trivers (2006) do not claim that selfish genetic elements have been a major force shaping the evolution of organism design. If anything, Burt and Trivers have catalogued the exceptions that prove the rule that genes are selfish (*sensu* Dawkins 1976), still viewing them as exceptions rather than changing the understanding of the rules in of themselves. In this regard, Burt and Trivers (2006) sits comfortably within the tradition that argues that, in the long run, the genome is the unit of selection, especially regardless of sexual reproduction. In the short run, selfish genetic elements may demonstrate the gene as the unit of selection, but in the long run it is as if there is a 'parliament of the genes' (Leigh 1971, 1977) that prevents any one gene from benefitting more than any other in the same genome. There is widespread support for this perspective, which extolls the importance of selfish genetic elements as a potentially major force in evolution, but with an actually minor effect (e.g. Alexander and Borgia 1978; Cosmides and Tooby 1981; Dawkins 1982; Rothstein and Barash 1983; Hurst et al. 1996; Haig 1997; Ridley 2000; Grafen 2006b; Strassmann and Queller 2010; West and Gardner 2013; Levin and Grafen 2019; Scott and West 2019). Even Dawkins (1982, p. 264) is included here because, whilst he maintains that genes are selfish, he also 'rediscovers the organism', where it is argued that most genes actually work for the good of the organism due to every gene benefitting from vertical transmission from parent to offspring. This is presented as explaining that basic observation that organisms are extraordinarily adapted, rather than design appearing at some other focal point in the hierarchy of biological organisation. Therefore, if 'The unity of the organism is an approximation' (Burt and Trivers 2006, p. 475), then it is widely accepted to be a very good one.

In recent years, there has been an attempt to reinforce the case for the unity of design. Hamilton (2001, pp. 331–332) may have endorsed this direction of work when he wrote that 'a genome that is perfectly "fair" to all kinds of fitness advantage, whether disruptive or not, couldn't be anything like what we would today call an organism—it could be a leaderless gang of DNA bits and pieces, a minute, weak-membraned protocell of the primordial ooze perhaps, nothing more'. Beyond rhetoric, Alan Grafen (2006b) brought the parliament of the genes into his Formal Darwinism project, which seeks to establish clear mathematical links between optimisation and inclusive fitness. Grafen (2006b, p. 556) assumed that the parliament of the genes is run by the largest coreplicon in the genome, which is taken to be those genes that only benefit from vertical transmission from parent to offspring. This assumption is critical to the connection between evolution by natural selection and the optimisation of an individual's inclusive fitness (as a singular value). West and Gardner (2013, p. 581) also invoked the parliament of the genes to dismiss the importance of selfish genetic elements to organism design, emphasising the mainstream view of their rarity. Thomas Scott and West (2019, p. 9) went further still to use modelling to argue that the 'debate over the validity of assuming individual level fitness maximisation has usually revolved around whether selfish genetic elements are common or rare' but 'even if selfish genetic elements are common, they will tend to be either weak and negligible, or suppressed'. Using meiotic drive as an example (amongst others), Scott and West (2019) explored the scope for a selfish genetic element to bias phenotypes, such as sex ratio. A key factor in the model was the comparison of the harm to the organism against the cost of suppression; if there is greater harm to the organism, a suppressor of the selfish genetic element is permitted to spread at higher costs of suppression. The results were explicitly argued to show mathematical support for the parliament of the genes, and consequently the view that whilst selfish genetic elements can exist, they are inconsequential to organism design. Through the reasoning behind the parliament of the genes, maladaptation caused by selfish genetic elements could equally be said to be either weak and negligible, or suppressed.

* * *

Whilst there are many points of agreement between the perspectives in *Genes in Conflict* and here on maladaptation, there is a major disagreement over the importance of selfish genetic elements to organism design in the long run, owing to a dispute over the logic of the parliament of the genes. The parliament of the genes appears to be the latest attempt to justify the Paleyan view that natural selection acts benevolently in the interests of the organism; although it admits that the short run can be manipulated by selfish genetic elements that 'break' an optimised trait, it asserts that organism design can be very well understood without any knowledge of them because design depends on the optimisation process over the long run. In defence of this perspective, it may well be argued that behavioural ecologists want to study the optimality of trait designs without getting waylaid by needing to understand the genetic basis of the phenotypes of interest (Grafen 1984). But such a defence has become increasingly problematic as, when the genetic basis of traits has been identified, it has often led to a rethink of the evolution of traits—including the moth colourations (Majerus 1998), sex chromosomes (Abbott et al. 2017), and greenbeards (Madgwick et al. 2019) already discussed. Here, again, there is the same peculiar juxtaposition of an argument that simultaneously asserts the importance of genes to design 'in theory', and yet also dismisses their importance 'in reality'. It is one thing to admit that an area of research like the genetic basis of traits is of little personal interest, but it is another to repeatedly suggest that it is actually unimportant to the evolution of traits; the latter claim is becoming untenable.

Of course, to the extent that is demonstrated by Scott and West (2019), the parliament of the genes works to make the unity of the organism a very good approximation. But: exactly what is that extent? The theoretical model convincingly shows three key results about trait distorters under various assumptions, which could include some forms of meiotic drive that are taken as the primary example. First, a meiotic driver is under selection to be more effective at biasing its transmission into the next generation. Second, the more effective the meiotic drive is at biasing its transmission, the more likely it is to be suppressed, on the basis that a more costly suppressor would be able to be favoured by natural selection. Third, if, as is to be expected, the meiotic driver is only a small fraction of the genome, natural selection tends to only favour a small extent of trait distortion (e.g. sex ratio bias) from this element in the long run. These points can be left undisputed because a major criticism is the connection between the extent of trait distortion and the fitness costs of drive and suppression. With a view to the inclusive fitness maximisation paradigm, the model focuses on the phenotypic consequences on the level of trait distortion to demonstrate the expectation that traits will be close to the genetically blind prediction of their adaptive optimum. This may be reasonable when considering that inclusive fitness is based on relative fitness—or even, the marginal increments in relative fitness attributable to a hypothetical gene for a trait. But it does not assess the fitness consequences of the genetic basis of the trait; the costs of drive and suppression can still impact the absolute fitness of organisms, and so play a major part in organism design. So, the model neglects to consider absolute fitness—or its consequences.

As an overarching criticism, the theory behind the 'parliament of the genes' can be contrasted with an alternative perspective on the genome as a 'society of genes' (taken from a chapter title in Ridley 1996, with its implications developed in Yanai and Lercher 2016, pp. 43–46). Whilst Dawkins (1976) emphasised that genes are selfish, in the sense that each has the potential to be favoured by natural selection only because of their ability to replicate (rather than their effects on organism fitness), Itai Yanai and Martin Lercher (2016, p. 259) emphasise that it is important to understand 'how interactions in the [genomic] society influence each gene's success'. Recognising the genome as a society shifts the focus of the attribution of success to genes from their selfish motivation to the diversity of actual relationships between genes (and organismal traits) that are favoured by natural selection. In this regard, it is essential to recognise that genes have specialised functional roles within the genome, like butchers, bakers, and candlestick-makers in human societies. Building beyond Yanai and Lercher's (2016) discussion with an understanding of maladaptation, the parliament of the genes could be likened to the assumption that each gene receives the same pay for their work, whereas the society of genes suggests that each may receive a payoff in accordance with the competitiveness of their trade. Some genetic loci may be more competitive than others because of the opportunities presented by their functional role in organism design. This does not necessarily bring genes into conflict in order to enforce the receipt of the same fitness. Instead, from a societal perspective, it is clear that cooperation in the genome does not need to be egalitarian to be successful. Just as evolution by natural selection does not act 'for the good of the species', neither does it act 'for the good of the genome'.

This is conspicuously the case for maladaptation (Madgwick 2020). An allele for maladaptation would be found at one of those more competitive loci, where there are opportunities to obtain higher relative fitness at the expense of the absolute fitness of alleles in the rest of the genome—and this can be in a way that cannot be suppressed, even at the extreme where that suppression is cost-free. Maladaptation requires negative relatedness, which is frequency dependent. The relative fitness gain of the allele for maladaptation is generally at its smallest when its frequency is lowest. As it spreads through the population, the relative fitness gain increases. Let us suppose that the genetic assortment of the recipients of the interaction that generates negative relatedness is specific to the locus for maladaptation, so there are asymmetric relative fitness effects that generate the potential for genetic conflict with other loci elsewhere in the genome. Were a suppressor to arise, even if the interaction had no correlation to the genetic assortment of the locus for maladaptation for the suppressor, there would still be genetic assortment of the third party that is not involved in the interaction. Consequently, through the ecological effect (or reproductive compensation in the context of meiotic drive), a potential suppressor may have a different relative fitness gain from the allele for maladaptation because of asymmetric frequency and assortment, but no capacity to gain from suppressing the allele for maladaptation (see the Appendix for a more rigorous demonstration).

Examples of selfish genetic elements—especially those that have persisted for a long time, like many of those that have been discussed (including greenbeards in Chapter 4)—provide evidence for the persistence of asymmetric payoffs in the society of genes. Such evidence may be viewed with some scepticism, if treated as a collection of oddities. So, as well as raising up an alternative theory, there is also a need to bring down the old one.

The basic problem with the parliament of the genes is that it relies upon the availability of suppressors. But genes are pleiotropically constrained in evolving to become suppressors because they also have other roles in the society of genes. A salient question is: how, within one or a few mutational steps, can a gene in the rest of the genome provide suppression? It is logical that most genes are going to be unable to suppress another gene because they produce functions that are completely unrelated; for instance, it would be far-fetched to suppose that a gene involved in bone development would be able to suppress meiotic drive. When suppressors have been found in nature, they tend to hail from genes functionally related to the one they are suppressing and be evolutionarily constrained by their costliness, such as the identified suppressors of the *t* haplotype (Gummere et al. 1986) that are very rare in natural populations (Ardlie and Silver 1996). This appears to be a common phenomenon; when attempts have been made to experimentally estimate general pleiotropic effects, little support has been found for the concept of 'universal pleiotropy', instead suggesting that genes are highly constrained to impact a few ancillary traits only (Wagner and Zhang 2011; see also Paaby and Rockman 2013). Rather than making suppression look inevitable, consideration of the functional scope of suppression and the best available data on pleiotropy may start to make the number of possible suppressors in the genome look vanishingly small. Whilst resistant alleles at the locus of a selfish genetic element have often been identified, suppressors at other locations in the genome are much rarer (Price et al. 2020). Even a cursory understanding of the molecular mechanisms involved can demonstrate why through the plausible complexity of suppressor evolution; for example, modelling has shown that a hypothetical meiotic driver with a poison–antidote mechanism may be invulnerable to full suppression of the poison and antidote regardless of when it is expressed, but be vulnerable to partial suppression of the poison only if it is not expressed during the sister-cell stage of meiosis (see Madgwick and Wolf 2021 for more details). Consequently, even among the functionally capable genes in the genome, there is no guarantee that they will act out their potential to be the right kind of suppressor, which will depend on the details of their mechanism.

Even if it is discovered that suppression can occur, it is another step to demonstrate that it occurs quickly and/or to a high level. Meiotic drivers are generally thought to evolve under very strong selection (Price et al. 2019). Of course, there may well be a detection bias, in that it is easier to identify meiotic drivers that rapidly spread with visible consequences for organism design, such as the *t* haplotype causing a shorter tail (Chesley and Dunn 1936). But resistance is often incomplete, like the reduction in the transmission bias of the *t* haplotype from over 90% to less than 65% (Gummere et al. 1986), which means that drive may still evolve but under weaker selection. The evolution of selfish genetic elements and their suppressors may be more like the out-of-equilibrium dynamics of an evolutionary arms race than the progressive movement to a final conclusion. For example, in red flour beetles *Tribolium castaneum* there is an allele for maternal effect dominant embryonic arrest (known as *Medea*) that leads to the death of offspring that do not carry the *Medea* element before hatching (Beeman et al. 1992). The geographical mapping of the distribution of the *Medea* element has shown that its frequency has varied widely across natural populations over 30 years of monitoring (Cash et al. 2019). Suppressors have been identified, but they have not spread through the population for reasons that are currently under investigation. As a result, it appears that there is much more to the dynamics of suppression than the 'parliament of the genes' neatly assumes.

There is still much to learn about selfish genetic elements and their suppressors, and one of the most fruitful areas of enquiry is likely to come using gene drives to solve grand challenges like pest control. In a phrase attributed to the physicist Richard Feynman, it has been said that 'what I cannot create, I do not understand'. This perspective is deeply challenging for biologists because organisms are such complex entities. Nonetheless, in recent years, the toolkit of genetic modification and the new discipline of synthetic biology have greatly expanded the creative scope, which could enable the driving of maladaptation into populations. It has long been suggested that a killer Y chromosome could be used to control mosquito populations to prevent malaria and other vector-borne diseases (Hickey and Craig 1966a; b), which was almost rolled out to control *Aedes* mosquitoes in the 1970s (Curtis et al. 1976). A killer Y chromosome has been shown to be able to drive population extinction in the laboratory (Lyttle 1977). Other laboratory populations did also evolve different mechanisms of incomplete resistance to the killer Y chromosome, which rescued their populations from extinction (Lyttle 1979; 1981). These seminal studies showed the potential for pest control using selfish genetic elements, which provides a means to study how populations respond to their release. Unfortunately, there was little appetite to take the particular technology to the field, due especially to safety concerns.

In more recent years, there has been renewed interest in using homing endonucleases to control *Anopheles* mosquitoes, especially to combat malaria (Burt 2003). Homing endonucleases provide the scope for a safe technology because they work by copying themselves into a defined location in the genome that can be directly specified in their DNA. Homing endonucleases have non-Mendelian inheritance, which converts a heterozygote for the homing endonuclease into a homozygote at a high rate, which may lead to their rapid spread through a population. If the location that the homing endonuclease inserts itself into is an essential gene, it can kill or sterilise the majority of offspring. Through homozygous lethality, the homing endonuclease would expect to reach a polymorphic equilibrium in the population. With a plausible engineered conversion rate of 90%, only 19% of zygotes may survive once the homing endonuclease has reached its equilibrium frequency. As mosquitoes are like most pests in having a very high rate of reproduction (and strong density-dependent effects), this is thought to have a limited impact on population size, which is also supported by the evidence from the insecticidal control of populations (see e.g. Charlwood 2020). The load on the population can be increased by increasing the conversion rate (Burt 2003); for example, a 99% conversion rate leads to 1.9% of zygotes surviving. Compatibly, it would also be possible to release multiple homing endonucleases at once, where the release of two homing endonucleases with a 90% conversion rate may also be able to reduce zygote survival to 1.9%. So far, multiple releases have proved to be the key way of achieving higher loads to suppress populations because of revealing constraints on improving the conversion rate (Deredec et al. 2008; Bull and Malik 2017; Unckless et al. 2017; Price et al. 2020).

Although it is early on in the study of artificial maladaptation, some important results have been uncovered. Homing endonucleases may not be as promising as they were initially thought because their mechanism of replication tends to be highly error-prone (Champer et al. 2017), although there may be some way to reduce this issue (Champer et al. 2018). Unfortunately, when errors do occur, they typically create a close to cost-free resistant allele that may rapidly outcompete the selfish allele, and so detract from introducing a load into the population at equilibrium. This outcome was not obvious from a genetically blind perspective that assumes suppressors or resistant alleles arise from random mutation. Other selfish genetic elements may well prove more useful at controlling mosquito populations. The cytoplasmic parasite *Wolbachia* has been proposed as an alternative method of pest control (Brelsfoard and Dobson 2009). As *Wolbachia* has been found in over half of all insect species (Hilgenboecker et al. 2008), it is potentially a highly versatile tool for population suppression. Further, *Wolbachia* have been implicated in several subpopulation extinction events in nature (e.g. Jiggins et al. 2002; Hornett et al. 2009). Field trials are beginning to show that *Wolbachia* can be an effective way to control mosquito populations and limit the spread of vector-borne diseases (e.g. Beebe et al. 2021, reviewed in Ogunlade et al. 2021). Indeed, one study estimated that dengue cases were reduced by over 95% after a release programme for *Wolbachia*-infected mosquitoes (Ogunlade et al. 2023). The ability of *Wolbachia* to evade suppression may relate to the details of its molecular mechanism (Hurst 1991). Going forward, the analysis of artificial maladaptations is likely to be hugely informative about the evolutionary dynamics of selfish genetic elements and their suppressors.

If the parliament of the genes has been incorrectly used to suggest a more powerful role for suppressors than in reality, then it should be possible to detect the costs to organism design in terms of absolute fitness. There are so many features of the way that organisms are designed that are difficult to understand, and it is all too easy to assume that such traits are adaptations for unknown reasons. An interesting but necessarily speculative example of this comes from the extraordinary difficulties that are encountered in human reproduction. For example, many humans embryos are naturally terminated at the start of pregnancy because of an abnormal number of chromosomes, which is known as aneuploidy (Hurst 2022). The rate of aneuploidy varies with maternal age, occurring on the typical range of between 20% and 40% of egg cells in European populations (Pellestor et al. 2006). Aneuploidy has often been thought about as a 'fact of life', where the complex machinery of human reproduction goes wrong for inexplicable reasons. But the explanation may lie with selfish centromeres. Eggs cells form through the division of cells, which in the final stages involves the formation of a large reproductive egg cell and a small nonreproductive polar body (that is eventually destroyed). From a maternal or paternal gene's perspective, half of the time it will end up in the egg cell and half of the time it is destroyed. Selfish centromeres are known to be able to bias this process, so that the selfish centromere ends up in the egg cell more often than not (Malik and Henikoff 2002).

To explain aneuploidy, an alternative mechanism to the usual one may be in operation (Hurst 2022). Consider a centromere that has a 50% chance of ending up in the egg cell or the polar body. After meiosis has taken place, if the centromere ends up in the polar body, it could detach from the meiotic spindle and reattach to the egg pole to create an aneuploid egg (here a triploid), which may then fail to develop properly and so be destroyed. Whilst this might appear to be disastrous to the centromere, it may indirectly benefit in a way that is analogous to spite among organisms; the centromere has ensured that the mother does not produce an offspring without the centromere, and has given itself a 50% chance of ending up in the egg cell at a future conception event. The logic assumes that the centromere may benefit because of reproductive compensation, where the early termination of the egg cell that does not carry the centromere enables an alternative egg cell that does carry the centromere to take its place next time around. Suggestively, reproductive compensation may be stronger in mammals than other vertebrates that lay clutches of eggs because a mammal is likely to rapidly conceive again, with much of the cost of raising an offspring coming after conception during pregnancy, breastfeeding, and ongoing parental care, which may explain why aneuploidy occurs at higher rates in mammals. What is seen in this example is the number of lines of evidence that need to be assembled together to substantiate a hypothesis of maladaptation, against the assumption of adaptation, which is just as insecure but nonetheless currently condoned without the need to accumulate such evidence.

More widely, perhaps one of the clearest demonstrations that there are costs from selfish genetic elements to organism design comes from those general mechanisms that prevent their spread. For instance, crossing over has been proposed to be a mechanism of meiotic drive prevention by breaking up unlinked genes (distributed throughout the genome) that work together to achieve drive (Haig and Grafen 1991; Hurst and Pomiankowski 1991; but see also Hurst and Randerson 2000). Such general mechanisms are imperfect, creating loopholes that can be exploited by specialised forms of meiotic driver, which is why meiotic drivers still exist despite the preventative role of crossing over. Interestingly, crossing over itself may also create new opportunities for meiotic drive, as seen in the case of Robertsonian translocations (Zanders and Unckless 2019). Beyond crossing over, there are also other general mechanisms that can suppress drive, including inbreeding (Bull 2017; Bull et al. 2019) and polyandry (Price et al. 2010), but each also creates new opportunities for maladaptation (in inbreeding depression and sexual conflict). Such considerations are often lacking in theoretical analyses because they are too complicated to bring into a mathematical model, but their effects can be crucial to the results. For example, a cytoplasmic male sterility gene could theoretically drive a homogeneous population to extinction (Charlesworth and Charlesworth 1979), but population structure (i.e. inbreeding from local mating) enables the selfish genetic element to persist at high frequencies in a viable population (McCauley and Taylor 1997). The prevalence of mechanisms like these is suggestive of the burden that selfish genetic elements place on organism design for many species, which may be why these preventative mechanisms are so ubiquitous across the diversity of life.

Beyond detection, there is also a great challenge in quantifying the costs to organism design. An interesting demonstration of this also comes from humans. Around 98% of the human genome is made up from noncoding DNA (i.e. DNA other than the exons of protein-coding regions), of which over 45% comes from transposons (see Yanai and

Lercher 2016, pp. 236–257 for a discussion). The ENCODE project famously showed that over 80% of the genome is transcribed or binds regulatory protein—or is otherwise associated with biochemical activity (Dunham et al. 2012), but it did not establish that any of the DNA (most of which is noncoding) is of any benefit to cells or the organism. It has become apparent that the vast majority of the genome is made up from repetitive elements produced by transposons that appear to serve no useful function. Almost 17% of the genome is made up from one such repetitive element called L1, which is a transposon of around 6,000 base pairs that can replicate itself within the genome (Deininger and Batzer 2002). There are over half a million copies of L1 in the human genome, most of which are defective pseudogenes that can no longer replicate themselves because of missing sections. It is comprehensible that L1 is so common in the genome as a selfish genetic element, but the persistence of a large quantity of defective pseudogenes is perhaps surprising because they are presumably costly without being of any use to any gene in the genome. What is perhaps even more surprising is that L1 is not alone. Alu is a shorter repetitive element of around 280 base pairs that accounts for just less than 11% of the genome (Deininger and Batzer 2002). If L1 is a parasite of the genome, then Alu is a parasite of L1. Although Alu is incapable of replicating itself, it can manipulate the machinery of the L1 transposon to get itself replicated within the genome. Both L1 and Alu are actively amplifying in the human genome. Many of the other L2 and associated small interspersed repeats that make up the rest of the noncoding DNA in the genome are retrograde transposons that were previously amplifying but are not longer active. Whilst this sounds grossly inefficient, and so likely to decrease absolute fitness, how can the effect on organism design be quantified?

One way that transposons might be costly is as a direct consequence of replication. It has been estimated that L1-mediated events account for around 0.2% of disease-causing mutations (Chen et al. 2005); 65 cases of L1-caused diseases have been identified (Goodier and Kazazian 2008). For example, the insertion an active form of L1 into the PDHX gene led to the deletion of 46,000 base pairs of the gene, which leads to sporadic pyruvate dehydrogenase complex deficiency that causes a progressive neurometabolic degeneration that is often fatal (Miné et al. 2006). But of course, transposons generate mutation-like events; most mutations are thought to be deleterious, but on the whole they are essential to fitness increase (e.g. Fisher 1930, pp. 38–41). In a similar way, Burt and Trivers (2006, p. 272) suggest that transposons may even be net beneficial from the occasional benefits they provide.

A potentially less ambiguous way that transposons are costly is through genome inflation in producing a large graveyard of noncoding DNA. Unfortunately, these costs are much more difficult to detect because of an inability to directly contrast uninflated and inflated genomes, but there are some species where comparisons can be made that hint at those costs. For example, in salamanders, it has been suggested that their swollen genomes, which are of the order of between ten and 100 times larger than the human genome, impede brain development (Roth et al. 1997). One consequence of a larger genome is a larger cell, which means that fewer neurons can fit into the same brain cavity. Salamanders have the simplest nervous system of any vertebrate, which is derived from a more complex ancestral state. Genome inflation may partly explain this unusual observation. Across salamanders (*Urodela*), genome size has also been shown to correlate with slower development (Pagel and Johnstone 1992) and lower species richness (Sclavi and Herrick 2019). Similar correlations have been drawn for other clades. Plants of conservation concern often have very large genomes; the genome size of threatened plant species is on average more than double that of nonthreatened species (Vinogradov 2003). A similar analysis of vertebrates found a more variable effect; reptiles and birds, which tend to have the smallest genomes, show that larger genome size correlates with higher risk of extinction—but there was no significant correlation in fish, amphibians, or mammals (Vinogradov 2004). Whilst it is difficult to precisely quantify the effects of genome inflation, there is little doubt that organisms with genomes that are of an inflated size because of a replicating transposon are unlikely to benefit from carrying such vast quantities of repetitive DNA.

Therefore, overall, whilst there is a broad consensus that the unity of organism design is an approximation, the critical question is: how much of an approximation? The current mainstream view, following the logic of the parliament of the genes, is that the unity of design is a very good approximation, as 'even if selfish genetic elements are common, they will tend to be either weak and negligible, or suppressed' (Scott and West 2019, p. 9). This position mistakenly assumes a greater role for suppressors than they have, especially in maladaptation, due to being pleiotropically constrained. Selfish genetic elements and their suppressors are in an ongoing arms race, and it is not clear how they will evolve, but it is likely to be highly costly to the absolute fitness of organisms. These costs can be very difficult to detect and quantify, but it seems the more that is discovered about selfish genetic elements, the more approximate the unity of design looks. With many examples of persistent selfish genetic elements that cause asymmetric payoffs throughout the genome, the egalitarian theory of the parliament of the genes (Leigh 1971, p. 245) should be reformed into the non-egalitarian theory of the society of genes (building on Yanai and Lercher 2016, pp. 43–46), where it is recognised that genes do not need to reap the same benefit to cooperate with one another in the production of organismal traits. In the long run, evolution by natural selection does not make genes act 'for the good of the genome', especially in maladaptation; instead, genes may produce traits that benefit their replication at the expense of individual fitness. The next chapter moves on from exploring specific examples of maladaptation to consider how more general traits, like some of those that have been discussed (e.g. sexual reproduction), give greater licence for the evolution of maladaptation.

6

Maladaptive transitions in complexity

Having critically evaluated the evidence for some specific examples of maladaptation from the social behaviour between individuals and (especially) within the body of a single individual, it is clearer that the best examples of maladaptation come from cases with clear variation that show that a trait need not be how it is. Now, the case builds to consider the maladaptive properties of deeper-seated traits, which are often shared by large taxa or even the entirety of extant life. The goal is to broaden the imagination to show that even though some features seem so fixed as to represent the best of compromises, they may also be otherwise. The focus, here, is not in the role of chance events or path-dependency in the evolution of life on earth, but rather the role of natural selection in producing maladaptations (although it is interesting to see how minor functional differences can alter the scope for evolutionary conflict, e.g. in human pregnancy; see Haig 2019). The case that is advanced is that maladaptation is part of the architecture of the natural world down to its very foundations, which is considered particularly through the nature of biological complexity.

The complexity of life is both a universal feature of all living things, and a graded feature that is widely accepted to be more extreme in some living things than others. Although there are no hard-and-fast definitions that seem to adequately quantify the various dimensions of sophistication that are found in living things, most biologists are willing to accept that whilst 'Complexity is hard to define or to measure, . . . there is surely some sense in which elephants and oak trees are more complex than bacteria' (Maynard Smith and Szathmáry 1995, p. 3). The driving question behind this chapter is a simple one: is such complexity adaptive, non-adaptive, or maladaptive? Most readers might immediately think that complexity is obviously adaptive because it fits the general understanding of the world around us where things are the way they are for a reason (Kohn 2004, pp. 12–16). To suggest that complexity could be maladaptive is a big claim that is antithetical and, as such, it is necessary first to establish a case for reasonable doubt from plain observations.

The intuitive case against complexity being adaptive rests on complexity being an ambiguous term. Sometimes complexity is used to refer to a result, like the solution to a difficult challenge. For example, mathematicians celebrate proofs that required extraordinary creativity to solve (Popper 1963, p. 241), from longstanding puzzles like Fermat's Last Theorem, through philosophically loaded conundrums like Godel's Incompleteness Theorem, and even on into the familiar rules like Pythagoras' theorem that forms the basis of trigonometry. Other times, complexity is used to refer to a method, which judges the pathway taken to the result. For example, mathematicians often prize simplicity above all else in their proofs, which has led to the bride's chair proof from Euclid being amongst the most popular of the four hundred or so proofs of Pythagoras' theorem. The more complex proofs of Pythagoras' theorem are still correct, but unnecessarily laborious.

So too in nature, complexity is ambiguous. An elephant with its vast size, sophisticated organs, and behavioural intelligence could be cast as a complex result from billions of years of cumulative natural selection, over which the original design has been increasingly perfected for survival and reproduction. On this basis, an elephant might be cast as better at survival and reproduction than its ancestors in deep evolutionary time, including those that looked much more like modern-day bacteria. Yet equally, an elephant is just as alive today as a bacterium like *Escherichia coli*, so: is it pursuing survival and reproduction by an unnecessarily complicated method? Whilst the existence of complex traits is suggestive of natural selection, it does not suggest whether such traits are adaptive or maladaptive, although it seems reasonable to suppose that observed complexity is often associated with adaptive fine-tuning. Alternatively, if the complex traits are unnecessarily complicated, this could imply that the trappings of complexity stem from evolution driving traits towards maladaptation, whereby survival and reproduction can become harder due to natural selection.

At first glance, it is not immediately clear whether the complexity of living things is an efficient result or an inefficient method, but there are two opposing perspectives that present starting points for investigation. There is a common-sense perspective that complex traits reflect the complex challenges of survival and reproduction. This notion can be found in Paley (1802, pp. 17–18) when discussing the famous inference of design for a watch found on a pebbled beach:

> every indication of contrivance, every manifestation of design, which existed in the watch, exists in the works of nature; with the difference, on the side of nature, of being greater or more, and that in a degree which exceeds all computation. I mean that the contrivances of nature surpass the contrivances of art, in the complexity, subtility, and curiosity of the mechanism; and still more, if possible, do they go beyond them in number and variety; yet, in a multitude of cases, are not less evidently mechanical, not less evidently contrivances, not less evidently accommodated to their end, or suited to their office, than are the most perfect productions of human ingenuity.

Darwin (1859, p. 84), making the same analogy to human designs, mirrors this sentiment: 'Can we wonder, then, that nature's productions should be far "truer" in character than man's productions; that they should be infinitely better adapted to the most complex conditions of life, and should plainly bear the stamp of far higher workmanship?'

Fisher (1930, p. 35) brought the Paleyan perspective into the foundations of modern evolutionary theory in the argument that natural selection increases mean fitness in the sense of lifetime reproductive success. An informative test of intuition comes from the Fisher's interpretation of quality–quantity trade-offs, where Fisher (1930, pp. 27–30) led the way in thinking that offspring quality should only be understood in units of potential quantity (see also Roff 2001). Indeed, species of organisms can even be classified as *r*-strategists or *K*-strategists on the basis of whether they increase their reproductive success by producing many low-quality offspring that are less likely to survive to reproduce for themselves, or a few high-quality offspring that are more likely to survive to reproduce (MacArthur and Wilson 1967, pp. 149–152). Paul Colinvaux (1980, p. 13) framed this in terms of efficiency, arguing that 'Having a few, large young, and then nursing them until they are big and strong, is the surest existing method of populating the future'; large animals, like elephants, have more efficient designs for survival and reproduction than bacteria like *E. coli*, by producing fewer offspring that are more likely to survive.

Fisher (1930, p. 38) places the organism as the method and result of evolution by natural selection. The organism expresses traits that alter its chance of survival and reproduction, and natural selection acts on those traits to preserve those individuals whose traits cause their greater survival and reproduction. From this perspective, perhaps the most remarkably complex trait is being able to reduce the essence of an organism, with all its sophisticated traits, down to the single-celled origin of the organism in the zygote—or even a tiny molecule of DNA that contains all the heritable information needed for an embryonic cell to recreate a new organism. An extraordinarily efficient result—unless, of course, the survival and reproduction of organisms is not the purpose of organism design.

Dawkins (1976) put forward the first clear case for an alternative perspective where modern organisms look more like an inefficient method, a throw-away vehicle for the passage of genes between the generations. Dawkins (1982, p. 249) developed this perspective further to dethrone organism survival and reproduction as the purpose of organism design because, as he saw it, 'Adaptations benefit the genetic replicators responsible for them, and only incidentally the individual organisms involved'. When gene replication is placed at centre-stage as the purpose of organism design, genes become something alien to the organism that can be helpful or harmful to its design, rather than an essential encapsulation of the information within it. Taking this to its logical conclusion, whilst accepting that 'Genes can also spread for what we think of as a more "legitimate" reason, say, because they improve the acuity of a hawk's eyesight', at the heart of this perspective is the idea that 'Genes will spread by reason of pure parasitic effectiveness, as in a virus' (Dawkins 1998, p. 304). Indeed, Dawkins (speaking in 2008 in conversation with Craig Venter; published in 2016, p. 199) argues that the genome is just 'a huge society of viruses'. Dawkins is intentionally subverting the perspective that puts organisms at the centre of evolutionary theory. Going beyond Dawkins's perspective, there are knock-on consequences that detract from the view that complex traits are a response to the complex challenges of survival and reproduction. Instead, for a selfish gene, the goal is simply to replicate itself, and producing an organism with thousands of other genes is an extraordinarily inefficient method that is superfluously complex for achieving this goal. Mark Ridley (2000, p. 8) explicitly identified this when he wrote that, 'If the business of life is to copy genes, most of our physiology suggests a loss of focus during evolution; it is not clearly relevant to replication'. The examples of selfish genetic elements demonstrate this loss of focus—and some even harm their organismal hosts whilst they do so. This perspective naturally lends itself to thinking that complexity is maladaptive.

There is much that could be debated in these two perspectives that lead to diametric assessments of the nature of complexity, but these debates quickly become more speculative than informative. A major issue is that measures like absolute fitness cannot be used to compare the adaptiveness of species because they conflate intrinsic and extrinsic (e.g. niche size) sources of fitness. Nonetheless, perhaps there are some informative examples where complexity changes over time in a way that appears independent of its extrinsic drivers. Here, humans provide an interesting window onto the problem through the complexity of peoples' lives within society changing on a relatively short timescale. The complexity is tightly connected with the increasing material prosperity of individuals, which is unequally distributed across society. There were many forerunning discussions (e.g. Galton 1869) about the connection between wealth and demographic infertility, but Fisher (1930, p. 253) was the first to point out that 'infertility is an important cause of wealth'. Shedding the eugenical overtones, Colinvaux (1980, p. 202) went on to provide an evolutionary ecologist's perspective on demographic transitions, suggesting a psychological explanation for the average changes in the number of children that people have in society over time through 'every couple choosing the number of children they think they can afford'. Colinvaux (1980, p. 202) suggested that 'their ambitions are such that their resources allow only a fraction more than the number that will replace them' because people have broad 'affluence expectations' that limit the number of children they think they can afford to have.

Consistent with such reasoning, a few countries like Japan have had a declining population because of increasingly low fertility, and many more countries have declining populations when excluding migration (UNDESA 2022). Over the coming decades, many more developed countries are projected to follow a similar population trajectory. Although many causal factors have been attributed by demographers at various levels of granularity, there is a reasonable case for grouping the vast majority of these together under the heading of 'affluence expectations' impacting how many children people believe they can afford. For example, one major social issue in the UK is the average house price compared to the average earnings, with 2021 house prices being 65 times higher than in 1970 but with average earnings only being 36 times higher (Borrett 2021; see also ONS 2023a), which makes it more difficult for young adults to accumulate the money to put down a deposit to buy a house. There is no requirement for a couple to own their house to have children, but it can be prohibitively expensive to rent a property with a room for each child, which reflects an affluence expectation. Moreover, the average age of mothers increased from 26.7 in 1970 to 30.7 in 2020 (ONS 2023b), which may reflect difficulties in obtaining an expected level of affluence. Increasing maternal age is also associated with corresponding increases in infertility and pregnancy complications, such as chromosomal abnormality (NCARDRS 2019). Consequently, through the lens of affluence expectations, despite technological advances and societal changes, declining populations may represent life becoming harder over the generations.

This is a necessarily speculative beginning to a discussion of the trappings of complexity being maladaptive, but it establishes a reasonable case for doubt. Complexity is often assumed to be the adaptive result of natural selection from the pressures to increase survival and reproduction in a challenging environment, but it could also be a maladaptive method that makes survival and reproduction more challenging in themselves. Having opened up the question of complexity's purpose, a more open-minded foundation is established from which to explore some more specific sources of maladaptation among the widely shared complex traits of living things.

* * *

When considering complexity across the diversity of life, it is apparent that some complex features are shared by elephants and *E. coli*, whilst others are common to both. This reflects the evolution of these traits in different lineages over vast spans of time, but these lineages also share a common ancestor. The first bacteria were around over two billion years ago, including the lineage that went onto to evolve into modern-day elephants. To understand whether complexity is adaptive, it is necessary to understand the context in which the different traits that underpin complexity were acquired.

Many different evolutionary biologists have proposed their list of the key innovations of life on earth, which roughly correspond to those complex traits that these authors believe to be transformative with respect to their special areas of interest. Andrew Knoll (2003) in *Life on a Young Planet* focuses on the evolution of the key traits of bacteria, like cell walls and aerobic respiration, which have large effects on the ecology of whole planet, by surveying their extant diversity. Richard Southwood (2003) in *The Story of Life* prefers to focus on the evolution of the main animal groups, especially with regards to the key changes in their anatomy from the fossil record. Dawkins and Wong (2004) in *The Ancestor's Tale* even more specifically focus on the evolution of the main animal groups, using their extant descendants to explore how life cycles have changed over time. Yet more specifically, Jared Diamond (1991) in *The Rise and Fall of the Third Chimpanzee* explores the key innovations in hominid evolution, often relying on archaeological evidence of the structures of human society. Whilst such sweeping reviews put different lists of innovations as the benchmark of evolution, all share the placement of complex traits as adaptations to new environmental challenges in chronological order over a gradual process of cumulative change. Consequently, although each author would make their case otherwise, the listed innovations are merely narrative devices for framing the discussion of a topic of subjective interest, rather than objectively classifying a drastic change in the complexity of the trait designs of life over evolutionary time. Therefore, the salient question is: what innovations matter for the special focus on complexity?

There is one list of key innovations that purports to directly address the traits that are most important for unlocking new patterns of complexity, which itself emerged gradually over time. Huxley (1912) explored the complexity of organisms across the diversity of extant life, providing the first discussion of the phenomenon of individuality. Individual organisms were framed hierarchically, with heterogeneous parts—like the organs of the body—working together to form functionally integrated wholes, which are known as individual organisms. Different organisms—for example unicellular and multicellular organisms—have different numbers of hierarchical levels of parts making up the individual organism. John Bonner (1974) described such extant diversity in more evolutionary terms, whereby multicellular individuals can be formed from the aggregation of unicellular individuals. But it was Leo Buss (1987, p. 4) that first articulated that 'Individuality is a derived character' and that 'the major features of evolution were shaped during periods of transition between units of selection [i.e. individualities]' (Buss 1987, p. 188). The focus of these early discussions was primarily on the phenotypic nature of the individual—especially whether it was unicellular or multicellular, which was the most obvious hierarchical feature of individuality—and on explaining the evolution of phenotypic complexity by natural selection (see also Bonner 1988).

John Maynard Smith (1988, p. 221), addressing the same problem of complexity, took an alternative approach to the notion of individuality, focusing on the genotypic nature of the individual. Whilst the previous descriptions had focused on the different number of phenotypic levels of hierarchical organisation among extant individualities, the focus on the genotypic levels gave rise to a more speculative hierarchy: loose replicators, compartmentalised replicators, bacterial cells, eukaryotic cells, multicellular organisms, social groups, species, and so on. The transitions between different organisations mean that, following Buss (1987, p. 4), it could similarly be said that the genome—or equally the entities that are referred to as a gene—is a derived character. Moreover, Maynard Smith (1988, p. 223) identified that there was a symmetry to the transitions between levels of organisation because, 'at each transition, selection for "selfishness" between entities at the lower level would tend to counteract the change'. Consequently, a transition was marked by the evolution of mechanisms that suppress competition at the lower level.

This conjecture was developed into 'the major evolutionary transitions' framework (Maynard Smith and Szathmáry 1995), which suggested an explanation for the evolution of complexity. Complexity was interpreted intuitively, but also imperfectly quantified using the number of DNA base pairs within an individual's genome, especially those regions that code for protein. Whilst this captures how elephants have more DNA than *E. coli*, there are many plants that have much larger genomes than elephants from duplications and hybridisations that do not drastically alter phenotypes, such as *Paris japonica* that has almost 40 times the number of DNA base pairs of an elephant. Regardless of its imperfections, as arguably any metric of complexity would have, this definition of complexity focuses attention on the information required to recreate an organism. Of course, as has been discussed, such complexity could be required for a more complex *result* in a more finely adapted organism, which is the intended implication, or it could capture a more complex *method* in a more inefficiently programmed organism. This is not directly addressed, presumably to avoid opening up a philosophical debate on the nature of biological complexity, which Maynard Smith (1988, p. 221) seemed particularly keen to avoid.

Focusing on complexity in informational terms, Maynard Smith and Szathmáry (1995, pp. 13–14) also describe transitions in terms of changes to the flow of information between the generations. Although the major phenotypic changes in individuality are included because they change the entity that is passing on its information to the next generation, there are also transitions that only have appreciable effects on the genotypic structures of individuals, such as sexual recombination and spoken language. These genotypic changes are still described in terms of 'higher level' organisations, presumably because even if they do not increase complexity in the intuitive sense of more information, they do increase complexity in the sense of having a more complicated structure to that information. Nonetheless, the mixed treatment of phenotypic and genotypic transitions together is perhaps surprising because they are of a different nature. Maynard Smith and Szathmáry (1995, p. 61) argued that the division between phenotype and genotype is, after all, only another major transition in the transmission of information between the generations, separating a store of information from the consequences of information extraction from it.

Queller (1997) further argued that both must be treated together because these transitions change the nature of competition, and so are a messy reality. The phenotypic and genotypic transitions are important for understanding where natural selection builds design in nature, wherein the individual organism necessarily is important because of how Queller (1997, pp. 187–188) argues that biologists define it: 'We designate something as an organism, not because it is n steps up on the ladder of life, but because it is a consolidated unit of design'; as such, 'Of course, organisms compete more than they cooperate; if they do not, then we simply shift our gaze upward until we discover the entity that fits the bill, and we call that the organism instead'. So, there is a circularity to wondering why competition occurs most strongly at the level of the individual organism because the identification of the organism relies upon identifying such competition. Instead, there is more to be gained from questioning why competition is weaker at lower or higher levels, and how this can change through natural selection.

The commonality, as Maynard Smith and Szathmáry (1995, pp. 6–7) emphasise, that all transitions share is a shift from entities that contain information replicating independently before a transition to replicating collectively after a transition. The basic advantage here is from a division of labour, where there are benefits from entities focusing on performing different tasks more efficiently, like the specialisation of the organs of the body. As Williams (1966, pp. 92–101) had already identified in the case against Wynne-Edwards' argument for group selection, natural selection could only favour a trait by selection at the level that the trait is expressed (see also Gardner 2013 for a discussion). So, the transition to replicating collectively must stem from the advantages of individual entities replicating together. After a transition, it is easy to imagine the advantages of individual entities replicating together because they generally have a shared fate through the same route into the next generation (*sensu* Dawkins 1982, p. 264). The challenge comes from any selfish advantages of replicating independently, which is why Maynard Smith (1988) originally suggested that a transition is marked by mechanisms that suppress the advantages of selfishness, especially during a transition (that has been expanded by Frank 2003). To evolve such mechanisms, natural selection must have a key role in transitions.

After a transition, Maynard Smith and Szathmáry (1995, p. 9) suggest that,

> If an entity has replicated as part of a larger whole for a long time, it may have lost the capacity for independent replication that it once had, for accidental reasons that have little to do with the selective forces that led to the evolution of the higher-level entity in the first place.

This phenomenon is described as 'contingent irreversibility', and it implies that the suppression of lower-level selfishness is unnecessary forever after a transition because entities may be unable to act selfishly anymore. Contingent irreversibility is as much an observed fact as anything else to the conception of transitions, and so it is acknowledged that irreversibility is a tendency rather than an impossibility. Selfishness can still arise at lower levels, but there may well be different mechanisms involved in the origin and maintenance of a higher-level entity (see also Bourke 2011, pp. 20–21). Indeed, in keeping with Dawkins (1982, p. 264), Maynard Smith and Szathmáry (1995, pp. 9–10) suggest that the maintenance of the higher-level entity may well rest on the logic of 'parliament of the genes' (Leigh 1971, p. 249) due to the incentives from being genomically organised (even if this is only an approximation). So whilst the origin of the higher-level entity may depend upon a specific adaptation for competition suppression, the maintenance of the higher-level entity may simply depend on the extent of the incentives that the new structure of information flow has created for the lower-level entities, rather than on the persistence of a costly mechanism of suppression.

One way or another, the major evolutionary transitions framework suggests that there is a direction to much of evolutionary change towards greater complexity, although it is by no means deterministic. The directionality arises from the compounding of lower-level entities to form higher-level entities into individual organisms that have greater complexity in the sense of multiple levels of hierarchical organisation. Of course, as Maynard Smith (1988, p. 221) was well aware, there is a 'zeroth'-level starting point, but the focus on the role of natural selection during transitions suggests that it is natural selection that drives the increase in complexity. Although it is difficult to ascribe adaptation when the entity that is the unit of design (*sensu* Queller 1997) keeps changing, the general advantages of division of labour create persistent selection for cooperation that indirectly favours greater complexity. Andrew Bourke (2011, pp. 27 and 180) has made the implications of this reasoning explicit, in framing transitions as adaptive events that increase the lower-level entities' fitness during and after transitions (see also Birch 2012, p. 575 for discussion). Moreover, contingent irreversibility means that once a transition has taken place, there are rarely the incentives to go back to a simpler mode of organisation (as a rule with exceptions for some lineages, especially parasites). Therefore, from its origins to its current form, the major transitions framework can be summarised as putting forward an adaptive explanation for the evolution of complexity through being favoured by natural selection.

The major evolutionary transitions framework is by no means the only available explanation of the evolution of complexity. Published shortly afterwards, Gould (1996, pp. 19–20) takes an anti-progressive view of the evolution of life on earth, reacting against claims that there is 'a tendency for life to increase in anatomical complexity, or neurological elaboration, or size and flexibility of behavioural repertoire, or any criterion obviously concocted (if we would only be honest and introspective enough about our motives) to place *Homo sapiens* atop a supposed heap'. Building on the basic premise that Maynard Smith (1988, p. 221) dismissed as 'boring', Gould (1996, p. 171) expands the case from Darwin that, given that 'Life had to begin next to the left wall of minimal complexity', the evolution of life only appears to have increased in complexity, giving 'the bell curve of complexity for all species a right skew, with capacity for increased skewing through time'. More directly attacking what he perceives to be the motivation behind arguing that there is the progressive increase in complexity over life's evolution, Gould (1996, p. 175) goes on to say

> If we could replay the game of life again and again, always starting at the left wall and expanding thereafter in diversity, we would get a right tail almost every time, but the inhabitants of this region of greatest complexity would be wildly and unpredictably different in each rendition—and the vast majority of replays would never produce (on a finite scale of a planet's lifetime) a creature with self-consciousness. Humans are here and by the luck of the draw, not the inevitability of life's direction or evolution's mechanisms.

Accordingly, Gould puts forward a nonadaptive explanation for the evolution of complexity. The appearance of complexity is the somewhat inevitable consequence of life's simple origins because life cannot get any simpler without not being life. Consequently, whilst it may appear that life has evolved to become more complex, this may have much more to do with constraints on complexity than the action of natural selection.

Going forward, in addition to the adaptive explanation of Maynard Smith (1988) and the non-adaptive explanation of Gould (1996), a maladaptive explanation of the evolution of complexity is proposed. As maladaptation has symmetric features to adaptation but with an opposite effect on individual fitness, whatever complexity means, it must still be favoured by natural selection for it to be maladaptive. Gould's (1996)—and indeed Maynard Smith's (1988)—points are fully accepted that not all lineages express the drive from natural selection to reap the advantages of a division of labour, and may indeed be driven in the opposite direction towards simplicity in some lineages. But Gould, labouring his argument against seeing humans as the pinnacle of evolutionary attainment too far, misses the precise claim in Maynard Smith (1988): that complexity within a lineage can hardly be the result of chance alone. Whatever the extent of complexity across life, as Paley (1802) highlighted, organisms really do appear designed for something in the fit between their traits and their environment, which must be the result of a non-random process that generates design. Indeed, on this basis, it becomes obvious that the general arguments of Gould (1996) and Maynard Smith (1988) are perfectly compatible but about different aspects of the evolution of complexity; Gould is focused on the evolved distribution of complexity across life, whereas Maynard Smith is focused on the evolution of complexity in some lineages. In keeping with Maynard Smith (1988), the goal is to explain not the distribution of extant complexity, but why it can increase. Therefore, leaving behind the nonadaptive explanation, the contention is not with the logic of major evolutionary transitions, but the accepted adaptive interpretation that natural selection must necessarily increase individual fitness during and after a transition.

* * *

Like the adaptive explanation of complexity, a maladaptive explanation of complexity must make use of common-sense reasoning to avoid unnecessary philosophical speculation about the definition of complexity. The contention is over the interpretation of complexity as an adaptive

result, rather than a maladaptive method. Although it would be difficult to say that every aspect of complexity is maladaptive, the deep-seated
traits underlying the phenotypic organisation may well express a maladaptive method because of their origins in major evolutionary transitions. The key features of the organisational complexity of living things
come from those major evolutionary transitions that involve the formation of a new individuality, where a group of individuals come to express
traits jointly and reproduce as a collective. This well describes the core
transitions from prokaryotic cells to eukaryotic cells, unicellular organisms to multicellular organisms, and social groups to eusocial groups,
which are all reasonably well-characterised evolutionary events (unlike
many of the other proposed major evolutionary transitions). The basic
problem in each case is that the new individuality is formed from entities that contain genes that were once capable of independent evolution
in their own right. As the new phenotypic organisation of the individual
evolves, the potential for maladaptation arises because the genes in those
old phenotypic organisations remain intact; indeed, the new phenotypic
organisation is only advantageous because some part of the old individuality remains functional in the new hierarchy. Therefore, the major
evolutionary transitions in complexity can be explained as maladaptive
because genes are used in the building blocks of new individualities,
despite those genes retaining their evolutionary potential.

A clear example of this comes from how the multicellular organism is
made up from many cells that are capable of independent reproduction,
which can lead to maladaptation. Multicellularity can be thought of as
an extraordinary feat of cooperation, requiring cells to be obedient to:
controls on their proliferation, regulation of their life or death, the enac-
tion of their specialised job, the greedless transportation of resources,
and the collective maintenance of the extracellular environment (Aktipis
et al. 2015). Whilst these restrictions may sound prohibitive, multicellularity represents a common solution to the challenge of growing larger (as
large cellular membranes can become unstable), having evolved at least 25
times in independent lineages (Grosberg and Strathmann 2007). Multicellularity can be exploited by parasites, which can be from distantly

or closely related species, and tumour cells, which are native body cells that evolve to be disobedient by mutation and selection. Many tumour cells are benign and/or successfully quelled by the immune system, but some go on to become malignant as cancers that can be fatal (Hanahan and Weinberg 2011). Fatality from cancer is particularly associated with animals rather than, say, plants because the movement of animal cells around the body provides a mechanism for a cancer to spread around the body; in plants, tumorous growths are reasonably common, but cells are statically constrained by their cell walls, and so such growths do not tend to disrupt the whole organism—and may even grow to the point of falling off under their own weight without long-term consequences for the functionality of the plant (Grosberg and Strathmann 2007). In this way, whilst there are different outcomes of cancer depending on the lineage, cancer is a general problem for all multicellular lifeforms. Indeed, the possibility of cancer is really but a general consequence of cells in the body containing genes that gives them the potential to reproduce independently.

The scope for independent reproduction is demonstrated most clearly in transmissible cancers, of which there are two contrasting natural examples in vertebrates. Tasmanian devil facial tumour disease (DFTD) is a newly arisen disease, first documented in 1996, caused by a transmissible, clonally reproducing cancer that spreads from one individual to another through facial biting, especially during mating interactions (Hamede et al. 2008). DFTD is highly malignant and tends to cause fatality within a year of infection, reducing the average longevity of wild devils from up to six years to less than three (McCallum et al. 2007). Since its origin, the devil population has suffered a massive population decline of > 90%, though other factors may also be partly to blame, and it is thought that DFTD could drive the species to extinction. It was initially thought that the spread of DFTD was because the low genetic diversity of the island population of devils prevents the immune recognition of transmissible cancer cells, but it is now thought that DFTD has traits that lead to it successfully evading the immune system, such as by downregulating the major histocompatibility complex factors (Pye et al. 2016).

Consequently, there is no reason to think that DFTD is exceptional to Tasmanian devils, and it is likely to be a persistent problem for the species—at least, without human interventions like vaccination (Belov 2012). The other naturally occurring transmissible cancer of vertebrates, canine transmissible venereal tumour (CTVT), is the oldest known malignant cell line on record, having been around for at least 6,000 years (Rebbeck et al. 2009); it is possible that CTVT is even older than this, but it is difficult to phylogenetically reconstruct its origin due to its clonal reproduction (Murgia et al. 2008). CTVT is much like DFTD in being a sexually transmitted disease that is adept at immune evasion, but it tends to regress in a matter of months after infection, and so does not come close to being fatal (Cohen 1985). It is hard to draw general expectations from but two examples. Even when including other examples such as in soft-shell claims (Metzger et al. 2015), it would appear that transmissible cancers can have various modes of infection, but there is no reason to think that they are a 'blip' or an otherwise short-lived phenomenon of little evolutionary consequence. Instead, transmissible cancers highlight the downsides for the organism of the genuine reproductive independence that is possible for somatic cells that would not otherwise pass on into the next generation of multicellular organisms.

With their reproductive independence, susceptibility to somatic cancer is arguably a maladaptation that is enabled by the major transition from unicellular to multicellular organisms. If this were the case, it is interesting to note that there is variation in the susceptibility of species to cancer. A famous observation, known as Peto's paradox, is that humans have thousands of times more cells than mice and live over thirty times longer, and yet do not have an appreciably higher risk of developing cancer (Tollis et al. 2017). Further, humans are by no means exceptional: there are many large, long-lived animals like elephants (or even whales) that have much lower rates of cancer than humans. Amazingly, a recent study concluded that the risk of developing cancer is largely independent of body size and organism longevity (Vincze et al. 2022). The explanation is likely to lie with natural selection (Boddy et al. 2015). Indeed, elephants have a risk of cancer that is two to five times lower than in humans, which could be largely attributed to having twenty duplicated copies of TP53—a well-known cancer-suppressing gene (Abegglen et al. 2015). The number of TP53 genes has increased over evolutionary time as mammal lineages have increased in body size, which has made cells more susceptible to programmed cell death with lower levels of DNA damage (Sulak et al. 2016). Despite such resistance, cancer is still known in elephants, as no mechanism has been discovered that 'solves' the problem of cancer that the evolution of multicellularity creates.

In this way, the evolution of multicellularity is a maladaptive transition in complexity in the sense that, whatever benefits it brings, it enhances the scope for maladaptation by making life harder. Quite simply, an *E. coli* bacterium need not worry about how many TP53 genes it has, or any of the large number of other immunoregulatory genes involved in cancer suppression. In the end, it could be that the balance of the benefits and costs of multicellularity are overall fitness-increasing because it helps more than it harms; but there is no reason that it could not also be the other way around. Such arguments may well rest on a case-by-case analysis of the benefits and costs of multicellularity to each species. But the general point stands: using reproductively competent entities as the building blocks of new phenotypic levels of organisation introduces the risk of maladaptation.

There are three important counterarguments to the suggestion that complexity is maladaptive that help to clarify the original argument. First, it could be argued that a transition would not evolve unless it increases fitness (*sensu* Bourke 2011) because, at the origin of the major evolutionary transition, there would be competition between the old phenotypic organisation and the new. It is a plausible counterargument, but mistaken for reasons that forerunning discussions have already identified. It is indisputable that major evolutionary transitions have their origins in the general advantages from the division of labour, so it seems more than likely that such transitions occur through adaptive traits that increase individual fitness. But the origin and maintenance of a transition (*sensu* Maynard Smith 1988) can present entirely separate evolutionary forces. A major evolutionary transition is a revolution in the phenotypic organisation of a lineage, and it can take a long time for the full effects of that revolution to play out. Consequently, as populations evolve over time, there is no reason to think that the old phenotypic organisation is ever in direct competition with every step on the way to the new organisation.

Consider the evolution of sexual reproduction. It is likely that sexual reproduction evolved in unicellular lifeforms over a billion years ago because of the straight-forward advantages that it provides in enabling the repair of double-stranded DNA damage, by the recruitment of a homologous chromosome that the damage can be corrected from (Michod 1995, pp. 8–9). The consequences of sex for life on Earth have been inestimable. Sexual reproduction paves the way for the evolution of males and females, elaborate mate signals, biological species, meiotic drive, and so on—as well as being associated with eukaryotic lineages. Regardless of their direct involvement, the steps that are taken after the evolution of sexual reproduction add to contingent irreversibility, making it more difficult for sexual species to revert to asexual reproduction— even if it would increase individual fitness in its current niche (as has been argued for some species; see Williams 1975). While reversion is technically feasible in some lineages through single mutations that lead to parthenogenesis, the resulting organisms are 'hopeful monsters' (*sensu* Goldschmidt 1929) with many non-adaptive traits that are designed for the habit of sex. For instance, in parthenogenetic lineages of whiptail lizards (*Cnemidophorus uniparens*), asexually reproducing individuals still seek, compete and undertake copulatory behaviours with each other even though this is not strictly necessary; it transpires that male-like stimulation is necessary for asexual reproduction because of behavioural controls on ovarian growth (Crews et al. 1986). Presumably, in their sexual ancestor, the behavioural controls were adaptive in avoiding wasting effort on ovarian growth without mating, but in asexual lineages it is superfluous. Consequently, in whiptail lizards, asexual individuals can only compete against sexual individuals (e.g. *C. gularis* or *inornatus*) in a biased way, which prevents evolution by natural selection testing the utility of sexual reproduction in the species' niche. In some lineages like mammals, the extent of the non-adaptation of the hopeful monster seemingly makes parthenogenesis impossible in that it has never been documented, which is a stricter kind of contingent irreversibility, but it really relies upon the same principles.

As such, there is no reason to think that a transition needs to evolve through steps that only increase individual fitness. Each lineage is alone in its niche, and consequently the extent that the old phenotypic organisation is in direct competition with the new organisation is limited to what can be achieved in a few mutational steps in a different direction. Consequently, as traits compound into the design of an organism, the trappings of complexity accumulate by natural selection on the available variation. There is no direct competition to test that all of those trappings are adaptive—or that the big traits like sex that lead to a diversity of new selection pressures are net adaptive.

The second counterargument to the suggestion that complexity is maladaptive is that a transition would only be successful if there is some property or mechanism that suppresses competition between the entities of the old phenotypic organisation (*sensu* Maynard Smith 1988). This counterargument mistakes a quantitative phenomenon with a qualitative one. There is a dangerous idea that can creep into biological thought that it is desirable to aim for conceptual neatness regardless of its reality (Mayr 1982, pp. 15–18). Of course, there can be properties or mechanisms that suppress competition, but they have an extent, which should not be misunderstood. A major evolutionary transition opens up new opportunities in the niche of a species, as well as creating new challenges in the scope for maladaptation. It is 'tidy' to think that natural selection drives a major evolutionary transition because of the advantages of a new phenotypic organisation, and then resolves the problems it creates by the evolution of a mechanism that suppresses competition in that new phenotypic organisation. But natural selection is not a rational designer that needs to close one door before it opens another. Instead, natural selection could be said to focus on driving whatever is of the greatest advantage. This can lead to the toleration of surprisingly high levels of competition between the entities of the old phenotypic organisation, alongside reaping the benefits of the new phenotypic organisation.

Consider the evolution of the eukaryotic cell. Over 900 million years ago, the eukaryotic cell is likely to have formed from the symbiotic fusion of the cells from different prokaryotic species (as originally proposed by Margulis 1970). As a result, some eukaryotic organelles retain derived copies of their ancestral genomes, such as the mitochondria that produce the ATP that powers cellular reactions. Whilst the presence of multiple genomes within a cell may sound innocuous, it provides the potential for genetic conflict between mitochondrial and nuclear genomes. Selfish mitochondria in budding yeast (*Saccharomyces cerevisiae*) have evolved that impose a fitness cost on growth rate as high as 30%, which can nonetheless spread through a population because the cost is outweighed by biased transmission into the next generation at rates of > 85% (versus 50% unbiased transmission; Jasmin and Zeyl 2014). The potential fitness loss from such selfish mitochondria is expected to plausibly reach high levels (Hastings 1999), and numerous mechanisms have been identified that can help to suppress selfish mitochondria including organelle selection, germline bottlenecks, and uniparental inheritance (Havird et al. 2019). These mechanisms can lead to their own problems; for instance, uniparental inheritance can lead the so-called mother's curse, where maternally inherited mitochondria are naturally selected to evolve traits that increase the fitness of females regardless of the costs that these impose on males (Ågren et al. 2019). So, suppression of one kind of conflict can open the door to new kinds of conflict.

The only seemingly definitive mechanism of suppression for selfish mitochondria is the transfer of genes to the nuclear genome. In comparison to the alpha-proteobacterial ancestor, it is likely that the vast majority of the original genome has been transferred to the nucleus, which may well have occurred because of the substantially lower mutation rate of nuclear genes (Roger et al. 2017). Often, the genes that have remained in the mitochondrial genome tend to have specific properties that seemingly prevent their nuclear transfer, such as producing highly hydrophobic or toxic proteins. There is, nonetheless, some variation in which genes have been transferred across species. The human mitochondrial genome contains close to 16,000 DNA base pairs with 37 identified (protein-coding) genes (Anderson et al. 1981), which is fairly typical of other animals, whilst the contrasting nuclear genome contains over 3 billion DNA base pairs and around 20,000 genes (Ezkurdia et al. 2014). Mitochondrial genome size varies from millions of DNA base pairs with over 50 genes in some angiosperms (Kitazaki and Kubo 2010) to just 6,000 DNA base pairs with three genes in the malaria parasite *Plasmodium falciparum* (Hikosaka et al. 2013)—although now there are examples of parasites that retain mitochondrial structures but have no mitochondrial genome (see e.g. John et al. 2019; Yahalomi et al. 2020). With no mitochondrial genome, the potential for selfish mitochondria is eliminated; though, it is important to say, nuclear genes may also cheat one another through mechanisms like meiotic drive when they are found in different locations of the nuclear genome. So, again, it would seem that suppression of one kind of conflict can open the door to new ones.

Since the origin of the first eukaryotic cell, it is interesting that the vast majority of modern-day eukaryotes have retained a mitochondrial genome despite the threat of selfish mitochondria and the vast amounts of evolutionary time. Indeed, the first eukaryotic cell is thought to be unicellular (Margulis 1970), but many modern-day eukaryotes have undergone another major evolutionary transition in individuality to form multicellular organisms, such as humans that retain their 16,000 DNA base pair mitochondrial genome (Anderson et al. 1981) despite this revolutionary phenotypic reorganisation. Moreover, some modern-day eukaryotes have gone even further with additional major evolutionary transition to form eusocial colonies, such as honey bees, which have a broadly similar mitochondrial genome to humans (Crozier and Crozier 1993), despite having undergone another phenotypic revolution. The widespread persistence of the mitochondrial genome over vast timespans demonstrates the extraordinary capacity of evolution by natural selection to tolerate competition between the entities of the old phenotypic organisation without resolving their conflicts.

The third counterargument to the suggestion that complexity is maladaptive is that natural selection can act on lineages through species selection, which would disfavour phenotypic organisations that tend to result in maladaptation. Species selection is a conflated and somewhat contentious idea. It is fairly well accepted that the speciation and extinction rates of lineages are related to the traits of individuals in those lineages (Jablonski 2008), which would presumably include maladaptation. For there to be a process akin to natural selection acting on the maladaptiveness of species, there is the additional supposition that speciation and extinction rates relate to maladaptation, so that those species that have individuals with the least maladaptation are also those that are more likely to diversify over evolutionary time. This may sound like a reasonable conjecture, given that maladaptation is expected to reduce population size, which could itself drive extinction or bring a population closer to stochastically driven extinction.

Accepting that species selection based on traits could occur for now, there is a major problem with the counterargument above. Species selection based on maladaptation relies upon the net speciation rate (i.e. speciation after extinction) being lower for species that are more prone to maladaptation. But those species may similarly be prone to adaptation. This is equivalent to how a species may have a higher rate of speciation and extinction. Consequently, species selection would be able to act not on a tendency to produce maladaptation in isolation, but rather on a tendency to produce more adaptation than maladaptation. This clarification is important in the context of major evolutionary transitions because it is indisputable that transitions have their origins in the general advantages from the division of labour, which is the adaptive rationale for the transition. But such transitions are cataclysmic events of organismal redesign that can bring with them an enormous scope for maladaptation. The trappings of complexity may still retain some of the benefits from the original adaptive advantage, but become dominated by the maladaptive disadvantages over time.

Consider the evolution of the eusocial colony. Western honey bees (*Apis mellifera*) are often presented as amongst the most perfect examples of a eusocial society, with the cooperative care of young, the reproductive division of labour, and overlapping generations (Wilson 1975b, pp. 430–433). Stingless bees in the genus *Melipona* are less perfect examples because of reproductive conflict, where between 5% and 16% of females develop into queens, whereas the colony optimum is thought to be < 0.02% from the expectation of honey bees; indeed, the closely related *Tetragonisca* stingless bees produce a similar number of queens to honey bees (Wenseleers and Ratnieks 2004). Further, the workers in *Melipona* colonies rapidly execute excess queens within 24 hours of their emergence from the cells, having wasted colony resources on their development up until that point (Wenseleers et al. 2004). It is as if the individuals that develop into queens have evolved to cheat eusociality like a cancer evolves to cheat multicellularity. An even more closely cancerlike example of a social parasite comes from honey bees, where an asexual strain of pseudoqueens (*Apis mellifera capensis*) reproduce parthenogenetically by automictic thelytoky (Martin et al. 2002a). This strain lays eggs despite the presence of a queen in the colony, and secretes a queenlike pheromone to avoid those eggs being eaten by workers exhibiting policing. Eventually, the levels of reproductive cheating overwhelm the resources of the colony resulting in its death. On human timescales, the asexual strain has invaded colonies of *Apis mellifera scutellata* in the north-eastern region of South Africa, causing the loss of all unmanaged colonies (Martin et al. 2002b). Amazingly, genetic analysis has suggested that the asexual strain has arisen by a single nucleotide difference (Aumer et al. 2019). Beyond cancer-like parasites, there are of course a wide diversity of distantly related social parasites, with thousands of species having been documented (Wilson and Hölldobler 1990, p. 436). With all these consequences of evolving eusociality, the question is begged: is a eusocial colony necessarily more efficient than a social colony? If eusocial colonies are so much more adaptive, why are more perfectly eusocial species so exceptionally rare among the wide diversity of social insects? If there were species selection, wouldn't these eusocial species be obviously outcompeting the 'socially inferior' species that are their ecological competitors?

The cases for and against species selection have previously been a matter of some—largely philosophical—debate. It has been persuasively argued at length that species selection is likely to be a weak force acting on organism design in comparison to natural selection on individuals within a species (Williams 1966, pp. 96–101). As Dawkins (1982, p. 106) carefully disentangled, 'a belief in the power of species selection to shape simple major [palaeontological] trends is not the same as a belief in its power to put together complex adaptations such as eyes and brains'. Whilst species may have traits that make them more or less liable to develop maladaptations and so go extinct, there is nothing that prevents natural selection within a species from driving traits towards maladaptation. Consequently, whatever paleontological patterns there are seem likely to emerge out of the natural selection on individuals within a species, rather than any process of meaningful species selection based on the competitive proliferation and extinction of species (*contra* Williams 1992, pp. 23–37). For instance, a natural selection on species would presume that the tendency to evolve maladaptation was a result of a species' heritable information, rather than its environment; the 'survivorship bias' (Wald 1943; see also Mangel and Samaniego 1984) in the traits of existing species might simply reflect the niches that are most profitable rather than the lineages that have adapted best. Beyond theoretical arguments, the diversity of organismal structures and the observable conflicts within organisms reinforce the view that species selection has a weak impact on organism design. This is perhaps unsurprising when taken in perspective: the assumption that species must evolve by a process that closely resembles the natural selection of individuals neglects the vast plurality of plausible alternative evolutionary mechanisms that had to be ruled out before natural selection became accepted as the driver of trait evolution (see Bowler 1983); there are still many other options for how species might evolve.

Overall, then, the emerging perception of organism design is one of neither exclusive adaptation nor maladaptation, but something in between. Major evolutionary transitions have an adaptive rationale in the division of labour. But this result is achieved through a mechanism whereby evolution by natural selection permits flaws to be built into the design of organisms that expand the risk of maladaptation. By using entities that were capable of independent reproduction as the building blocks of new phenotypic organisations, relying upon them retaining the genes that give them evolutionary potential, major evolutionary transitions exhibit a maladaptive method. Despite vast spans of time, natural selection has proved unable to prevent maladaptation from ever evolving, as even 'preventative' traits seem to have their own flaws that can be extensively exploited. Consequently, natural selection shows an extraordinary toleration of competition between the entities of the old phenotypic organisation that can drive maladaptation. There is no reason to think that the evolution of life on Earth has shown any clearly progressive patterns around the scope of maladaptation. Next, it is considered whether maladaptation is peculiar to life on Earth or whether it is to be expected of all life anywhere in the universe, by delving into the strange relationship between genes and organisms.

7

Deep origins of maladaptation

Some traits bear the hallmarks of maladaptation, and some of those are more general to all life, so can it be implied that all life can potentially suffer maladaptation? Was maladaptation assured from the start, if it were possible to 'rerun the tape of life' (*sensu* Gould 1989, p. 283) on Earth? Or should it be expected that life elsewhere in the universe may take a different course and not exhibit maladaptation? These are deep questions, and it is difficult to know that there is a source of evidence that can help to sift through answers. Most obviously, inferences can be made based on the patterns of extant life, but this is problematic. Such inferences would be relying upon the snapshot of life on Earth as it is now, which makes the inferences biased by the global environment as it currently is. To get around this, the fossil record can be used to look back in time to extinct life. But maladaptation often arises from behaviour, which is difficult to infer from the morphological remains or impressions that are left behind. Further, there is still the problem of common ancestry, as life on Earth has evolved from the proliferation of successful individuals. Consequently, many living things share features because they evolved them in a common ancestor, rather than because that feature arose independently because of an evolutionary tendency. Taking this idea to its logical extreme, given that all life is thought to have evolved from a universal ancestor, relying on evidence from extant or extinct life on Earth could be like drawing a trend from a single data point. The only way to overcome such a 'survivorship bias' (Wald 1943; see also Mangel and Samaniego 1984) would be to gather data points from living things with independent origin events, such as from life on other planets. But, unfortunately, if life does exist elsewhere in the universe, it currently lies beyond detection. Therefore, it would appear that there is no evidence to test speculation to address these questions.

There is one last source of hope: theory. Based on current knowledge, theoretical expectations about what cannot be easily tested can be used to arrive at reasonable speculation. This may sound like a weak line of evidence, not least because it is reluctantly being proposed after having exhausted all empirical lines of inquiry. But it may not be as weak as it might seem. For example, as Chapter 2 outlined, the key tenets of the mathematical theory of natural selection in population genetics were established prior to the discovery of the mechanism of inheritance from experimental work, using modelling to rigorously test the self-consistency of proposed mechanisms of inheritance that lay beyond the tools of observation at the time—and, indeed, very little changed following the discovery of DNA. So, here, theory can be used in a similar way to define the scope of reasonable speculation.

To address the ubiquity of maladaptation, the explanation for the appearance of design in nature can be leaned on. Design is both the fundamental hallmark of living things, and also the only way that life may continue to exist as environments change. Design only comes about through evolution by natural selection. A great deal of evidence has amassed on how natural selection works on extant life forms. Consequently, there are expectations of what is necessary for life to evolve by natural selection. In theory, these expectations can be probed by stripping away features of models until removing something breaks evolution by natural selection. In reality, this can be imagined as projecting theory deeper back in time, stripping away the modern-day complexity of life on Earth to arrive at the necessary features for exhibiting design as environments change. The deeper back in time that can be looked in order to reach simpler life forms, the more confident the speculation on the general properties of life, as those features of chance and necessity that have shaped life on Earth are removed. Going back into the depths of evolutionary history, at base there must somewhere be those first steps of biochemistry at the origin of life on Earth. This event, albeit very obscure as it is, is probably the limit of reasonably constrained speculation because, at least, it is known that it must have occurred. Therefore, to best understand the relationship between life and maladaptation anywhere in the universe, it can be asked: was the evolution of maladaptation assured from the origin of life on Earth? If maladaptation is an expected feature of those simplest life forms, then it is reasonable to think that it would also be an expected feature of all life anywhere in the universe.

At the outset, it must be acknowledged that there is a tendency among biologists to view speculation on the origin of life as little more than a seductive waste of time. Darwin (1859) gave precedent when he refused to be drawn on the topic in *The Origin of Species*, and in a letter to Joseph Hooker in 1863 was explicit that he thought 'it is mere rubbish, thinking at present on the origin of life' (Darwin 1887a, p. 18). Since Darwin, many eminent biologists from Haldane (1929; republished 1968) to Jacques Monod (1970) have felt that the absence of an agreed description of the origin of life is a major failure of evolutionary theory, more often to silence philosophical or theological gap-filling by critics than for any other reason. For this very reason, Darwin eventually succumbed to temptation in a famous letter to the same correspondent a few years later in 1871, discussing the potential for life to have begun in 'warm little ponds' (Darwin, 1887a, p. 18). Against the need for such hypothesising, it has always seemed evident to me that the origin of life is a question of chemistry, not biology (cf. Crick 1981, p. 37). Not only that, it is also a somewhat disreputable branch of chemistry because it is caught between the certainty that life did originate and the seemingly perennial lack of imagination about the possible forerunning mechanisms of life. Indeed, the extent of the possible diversity of life forms on Earth has continued to surprise scientists even into the modern era, with the discovery of abundant life in the deep ocean around hydrothermal vents in 1977 challenging the notion that all life on Earth ultimately depends on photosynthetic metabolism (Corliss et al. 1979). There are still just so many unknowns! Not that hypothesising cannot be entertaining, as almost every evolutionary biologist seems to have a preferred 'pet theory' that inevitably pulls together the particular threads of life's mechanism that are of personal interest.

So, alongside stating the question, it is also necessary to state what sort of evidence is being sought, from which arises the one biologically respectable purpose that there seems to be for discussing the origin of life. Ever since Darwin, biologists have been remiss for holding certain aspects of how life works as 'essential' because of this anachronism in the face of the overwhelming evidence of life's ability to transform itself. Firm boundaries describing when life originated (or indeed how to define life) definitely belong to this erroneous essentialism, but considering 'simple' systems can cut through the complexity of modern biology to bring its important features into sharper focus. A good example of this is Dawkins's (1976, pp. 14–19) discussion of the origin of life to distinguish replicators and vehicles, to draw out the importance of the selfish gene to the design of survival machines. As such, when discussed with discipline, speculation on pre-biotic evolution can help to more clearly see the fundamental mechanism of biological evolution at work all around us. This is not a banal conjecture of what chemicals were involved, but a redress for the theory of how evolutionary change takes place.

The most widely accepted hypothesis for the origin of life on Earth, in as far as biologists care, is something akin to an 'improbable soup', which was first popularised by Monod (1970), relying upon the idea that various simple biochemicals can be abiotically synthesised. The basic premise is that, at some point, just the right chemicals come together to form a sufficiently complex life form that can use environmental chemicals to make more of itself. As becomes increasingly apparent, there is much debate about the nature of this life, which all too often returns to the specific chemistry that it is thought to be based on. Nevertheless, once it has arisen, the core hypothesis would suppose that natural selection can then bridge the gap between these relatively simple life forms and their modern counterparts. In this way, the improbable soup requires two great suspensions of disbelief: the first in the jump (or saltation) from simple chemicals to a complex life form, and the second in the jump from the simple way of life of this life form to its modern equivalents. Proponents of the improbable soup have laboured to suggest how one of these jumps can be not as unlikely as it may initially seem, but all too often this plainly makes the other jump more drastically improbable. As a case in point, Monod (1970, p. 137) accepted that 'At present we have no justification for either asserting or denying that life made only one single appearance on earth, and that, as a consequence, before it appeared its chances of occurring were almost nil', but then took the power of natural selection for granted (Monod 1970, p. 134) to overcome the second jump.

Patently, any reasoning about an improbable soup suffers from a non-evolutionary way of thinking, and a few evolutionary biologists have tried to change preconceptions, most notably Cairns-Smith (1982). Making the case against an origin of life based on chance alone, Cairns-Smith (1985, p. 47) adroitly summarised: 'there was not enough time, and not enough world'. Monod (1970, p. 137) might retort that this belief merely stems from the way that 'It offends our very human tendency to believe that everything in the world is necessary, and rooted in the very beginning of things'. But the final word belongs to Cairns-Smith (1985, p. 7): there is no place for a 'scientifically respectable version of a special miracle' in evolutionary theory—even at life's origin—because it denies the reality of chemical evolution. When properly accounted, no great suspensions of disbelief should be required to understand the origin of life.

The required explanation, then, is not so much the discovery of the chemistry of the first life forms on Earth, as the origin of the kind of natural selection that drives biological evolution. Lots of chemical systems could be described as if they were evolving by natural selection, but it is debatable whether this description really helps to understand those systems (which may be better served through the descriptions of chemical reactions). For example, it is well accepted that the environment can 'select' in ways very unlike biological evolution by natural selection, such as favouring entities based on extrinsic features, which has been summed up as the 'survival of anybody' or 'survival of the first' (Michod 1999, pp. 26–27). On the other hand, natural selection is indispensable for understanding biological evolution, which is based on the selection of entities based on their intrinsic qualities in the 'survival of the fittest'. This is not to say that biological systems are not chemical in nature, but to set apart the concepts that define different kinds of change.

The results of biological evolution sit apart from its chemical counterparts. Biological evolution appears to be extraordinarily flexible, leading organisms to accumulate traits that appear designed, which allow them to exploit a wide range of environments. There is even an extent to which the chemical underpinnings of those traits are irrelevant to understanding what would evolve; indeed, neither Darwin nor Mendel were aware of the chemistry of life when deciphering the key pillars of evolutionary biology. As such, evolution often converges on similar designs by different means in distantly related species in response to similar environments, such as morphological adaptations of cacti and euphorbia to save water on different continents (Alvarado-Cárdenas et al. 2013). Further, biological evolution by natural selection enables living things to continue to evolve new designs as environments change, rather than being drawn into a dead equilibrium. For this reason, there is a superficial sense in which life appears to fight against the physicochemical laws of the universe—evolving fur against the cold, flight against gravity, and migration against seasonality. Whereas, chemical systems appear much more limited in this capacity to respond counteractively, tending to move in a direction of a gradual decay to homogeneity with their environment (which could be described as a natural selection, but clearly in a very different sense to the natural selection on biological systems).

Consequently, it is necessary to be clear about what distinguishes the open-ended evolution by natural selection of biological systems from the comparatively close-ended evolution of chemical systems. To this end, whilst trying to avoid overstepping the limits of constructive speculation, the intention here is to discuss the origin of life in order to piece together how chemical systems became biological, and to discern what this has to say about the prospects of maladaptation being inextricably linked with all life anywhere in the universe. The place to start is by considering the origin of life.

* * *

Of the preeminent theories of the origin of life, there is a natural division between those that posit that the organism came first and those that suppose that the replicator did so (Fry 1999). Organism-first theories are more often described as metabolism-first theories (as in Fry 1999), centring on the description of a method for the extraction of energy from the environment, and perhaps unsurprisingly this has attracted the attention of most physicists and chemists that have approached the topic. The distinction between metabolism and organism is of limited importance in this context because, either way, these theories interpret the origin of life as the origin of the first cells, which are the fundamental units of both production and reproduction. By contrast, replicator-first theories centre on the emergence of an entity that is capable of making copies of itself, which has mostly attracted the attention of mainstream biologists. These theories more often focus on a simpler origin through the emergence of a single molecule that is capable of catalysing its own production. Such an autocatalyst is thought to represent the origin of life as a precursor of a modern-day gene, which is the fundamental unit of inheritance and function. Nonetheless, it is indisputable that life as it is known in the modern day is dependent on cells. Squabbles over the definition of life quickly becomes tedious, as Dawkins (1976, p. 18) remarked: 'Human suffering has been caused because too many of us cannot grasp that words are only tools for our use, and that the mere presence in the dictionary of a word like "living" does not mean that it necessarily has to refer to something definite in the real world'. Consequently, setting aside how life is defined, there is the potential for both to offer insights about the origin of biological evolution by natural selection.

Alexander Oparin (1924) is often cited as the original author of a metabolism-first theory of the origin of life, writing in Russian with an English translation arriving in 1967. He was primarily concerned with specific speculations around the chemical nature of the primordial ocean and atmosphere (Oparin 1967, pp. 226–234). Although Oparin did not identify a clear 'origin of life' event, his theory was organism-first because he was interested in the properties of colloid chemicals in forming cell-like entities when specific sugars or proteins reach high enough concentrations (Oparin 1967, pp. 229–230). These proto-cells can have selectively permeable membranes much like modern cells, allowing some molecules to diffuse into and/or out of them whilst blocking others. Potentially, if just the right proteins were absorbed into a proto-cell that acted like enzymes to catalyse adding new units to the membrane, the cell could reach a critical size and then spontaneously divide in two, as an act like reproduction—and without anything like a replicator's involvement. Oparin (1967, pp. 230–232) speculated that those colloidal units that are better at reproducing would naturally outcompete those that are worse at reproducing, in a manner akin to natural selection. To do this, the colloidal units would need to bear qualities that are suited to the surrounding environment, leading to a kind of fit with the environment like adaptation.

Although the consensus about the details of the chemistry of the primordial ocean and atmosphere has changed (Fry 1999, pp. 113–117), many of these themes continue to be important in other organism-first theories to this day. Much of the early interest in ideas like Oparin's was critically tied to the chemistry (and/or physics) in question. Indeed, early support seemed to come from Stanley Miller's (1953) famous experiments in generating complex biochemicals in a laboratory setup that was supposed to mimic the primordial Earth (see also Orgel and Miller 1974). Melvin Calvin (1969, pp. 170–176) also pointed out that many of the structures of proteins that are found in modern-day bacteria are the consequences of protein-modification by other proteins—without direct specification from protein-coding DNA. In this way, Calvin suggested that there may be scope for proteins to modify themselves, including cutting themselves into pieces to form 'seeds' that could continue to grow and cleave off new proteins, again in a growth-like process of reproduction that is unlike replication. Moreover Sydney Fox (1980) had some success in forming proto-cells from reasonably simple proteins. Building on these findings, but taking these ideas away from the underlying chemistry, Stuart Kauffman (1993, pp. 298–312) used modelling to argue that complex sets of self-organising proteins could collectively reproduce themselves without there being anything akin to a genome that specifies how to code the proteins. All of these ideas point towards a protein-based metabolism that could take place within a cell-like entity without the presence of (or need for) any entity that could be described as a replicator in the pattern of modern-day genes.

Unfortunately, as was discussed well before much of this work (Horowitz 1959), it was well known that collective reproduction systems have a limited capacity for heredity because they struggle to reproduce modifications. In brief, the basic problem can be conceptualised by recognising that a random mutation that causes a change in one component is highly unlikely to be faithfully reproduced by the others because it is only indirectly responsible for its own reproduction (Eigen 1971; see also Maynard Smith and Szathmáry 1995, p. 71). Consequently, it is more likely that the evolution of such proto-cells would be driven by random drift rather than natural selection (Dyson 1982).

The original replicator-first theory is usually attributed to Haldane (1929), who originally published his ideas in the *Rationalist Annual*, which has subsequently been republished in collected works (e.g. Haldane 1968). Unknown to himself, Haldane had similar ideas to Oparin about the primordial environment (Haldane 1968, pp. 6–8), but a key difference emerged in that Haldane focused on the act of self-replication (Haldane 1968, p. 8). He considered the first molecule that makes more copies of itself to be something like a virus, arising in a hospitable environment (rather than a modern-day virus that inhabits an inhospitable host organism). As the environment changes, the self-replicating molecule may then evolve by natural selection to acquire the metabolic apparatus of the cell as a means of adapting to thrive in a harsher environment (Haldane 1968, p. 8). Haldane was vague about exactly when such an entity might be described as 'alive' (see also Haldane 1968, p. 4), but saw no issue in assuming its gradual ascension in complexity through evolution by natural selection from its simple chemical origins (Haldane 1968, pp. 8–10). For Haldane, then, the origin of life was extremely unlikely, and he supported the hypothesis that life had a single origin on Earth with evidence from the homochirality of life's modern biochemistry (Haldane 1968, p. 9).

Inspired by Haldane's thought experiment, some important experiments went on to characterise the evolutionary pressure from natural selection on the first replicators. An early study explored the evolution of a sequence of bacteriophage RNA in a hospitable solution in the presence of a replicase enzyme that would reproduce its RNA code (Spiegelman 1967). After almost 80 generations, the original bacteriophage RNA had degraded from a diverse code of 4,500 nucleotides to a repetitive code of around 220 nucleotides. The final sequence was 15 times faster than the original at replicating, with a six-fold increase in replication time from the smaller sequence and a 2.5-fold increase per unit of length from using the free nucleotides that were more abundant in the hospitable solution (Spiegelman 1971). In contrast to this top-down approach that begins from a modern-day starting point, a similar study was conducted from a bottom-up perspective (Biebricher et al. 1981). Starting with just a lone replicase enzyme in a hospitable solution, a simple RNA evolved that seemed to be limited to around 120 nucleotides in length (Eigen 1992). The results did not show a gradual evolution to a single RNA, but rather a population of similar sequences that persist under an error-prone mechanism of replication. These results suggest that there are constraints that limit the scope of the evolution by natural selection on these early replicators, preventing the runaway evolution of greater complexity with the mere origin of an autocatalyst (Ridley 2000). It is as if the replicators appear to be selected more for their replicability than for their functionality.

In response to the way that both organism- and replicator-first theories ignore the role of the other entity in shaping evolution, there are a few of conciliatory theories that hybridise elements of each. Interestingly, these theories often stem from a dissatisfaction that natural selection can straightforwardly bridge the gap between the simplest entities of interest and their modern-day equivalents. Indeed, Cairns-Smith (1985, p. 37) describes chemical evolution as 'up against the cliff-edge' of biological evolution, which is just so much more complex. Recognizing this problem led to the influential hypothesis of genetic takeover (Cairns-Smith 1982), where simple replicators are replaced by more complex replicators in what could be presumed to be a series of transitions. Indeed, whilst Cairns-Smith was primarily concerned with the original transition to the first replicators from crystalline minerals like clay, others have elaborated on the potential for the later steps to thioester, RNA, and then DNA replicators (see De Duve 1991). The lack of modern-day intermediates from these transitions is explained using the metaphor of higher technologies, which are more efficient at scale (Cairns-Smith 1985, p. 61) and better suited in a more competitive environment (Cairns-Smith 1985, p. 90), not least because high-technology predators would be easily able to pick off low-technology prey (Cairns-Smith 1985, p. 100). The result is that even the minimum of modern-day biological complexity appears to be completely separate from chemical systems. But Cairns-Smith provides an explanation for how they link up.

Cairns-Smith relies heavily on imprecise 'sketches' to explain the pattern of takeover events, which consequently are left rather vaguely defined. He defends that 'genetic takeover is meant to be a background hypothesis. It is meant to be vague—in the sense that it is a general frame within which to make more specific speculations' (Cairns-Smith 1982, p. 131), which he makes with reference to the chemistry of the first replicators more so than the evolutionary mechanisms at work during a genetic takeover. Nonetheless, a takeover could be summarised in three stages, following Cairns-Smith's (1982, p. 120) stage-less description of a takeover by some secondary genes from primary genes.

In the first stage, a new type of replicator arises and 'The secondary genes at first assist the primary genes in a minor way' (Cairns-Smith 1982, p. 120). Presumably, the new replicator necessarily has to be replicated to fulfil its function, which is why the old replicators used the new replicator as a means of producing useful traits. In the second stage, the primary genes 'take over the means of production of phenotypes' (Cairns-Smith 1982, p. 120). This is probably the most critical but also unclear stage of Cairns-Smith's description because he thinks it is obvious that the secondary genes represent a higher technology that is more efficient at producing phenotypes (which is why the primary genes used them in the first place). But this does not arrest the means of the production away from the primary genes, it merely adds a new means of production via the secondary genes. Interjecting some unstated reasoning here, it seems likely that, if most of the variation between individuals is driven by differences in the secondary genes, the stronger selection acting on the secondary genes would override the weaker selection on the primary genes through the phenomenon of selective interference (Madgwick and Kanitz 2021). Although this would not happen all at once, the stronger selection on the secondary genes could increasingly introduce randomness into the direction of change for the primary genes, which is thought to occur among modern genes as well (Gillespie 2000). The third stage involves the abolition of the primary genes as replicators because the means of production 'have become transformed in the meantime to the exclusive service of the secondary genes' (Cairns-Smith 1982, p. 120). Again, Cairns-Smith is not clear how this could happen, but presumably if selective interference makes the primary genes subject to random drift, then over time evolution will lead to the fixation of the primary genes, which rules them out as units of heredity. The process may also involve a direct component of natural selection, as the system may move to conserve the primary genes as a background for the evolution of the secondary genes (because mutations in the primary genes may become increasingly deleterious as the secondary genes evolve more rapidly to become more intimately co-adapted with them). As Cairns-Smith is more concerned with, the biochemicals that were the primary genes may still persist as units of function, like he argues proteins may have done, but they are no longer units of heredity.

It is for this reason that, whilst Cairns-Smith definitely focuses on the evolution of replicators, it would be an oversight to cast his ideas as an exclusively replicator-first theory. Cairns-Smith is not describing just the evolution of replicators, but the evolution of the complex biochemical processes that are going on all around them as well. Indeed, by suggesting that proteins were originally replicators, he explicitly addresses how metabolism and cells fit into his account of the origin of life, as the useful remnants of past takeover events. Further, Cairns-Smith (1982, p. 297) is insistent that the replicator is only a cue of the traits that it causes; the modern-day genome is not a blueprint for the organism, but rather contains information for how to modify the organismal machinery as it is received to bring about a trait. The replicator is not the only driver of evolutionary change in this story, with a significant role for the co-evolution of the rest of the metabolic chemistry. The key point is that, for Cairns-Smith, the complexity of metabolism did not arise all at once in an improbable soup, but arose through a gradual process of genetic takeover, where the biochemistry of metabolism came as a consequence of a turnover of different biochemistries. A key analogy that is used to explain this situation is the formation of an arch (Cairns-Smith 1982, pp. 95–99); although it appears to be miraculously held aloft when it is finished, with each stone supporting each of the others, it was created with scaffolding that holds up the arch during the intermediate steps whereby the arch is built up by stones that cannot support one another. And it is this analogy that has been the focal point of criticism because, by Cairns-Smith's (1985, pp. 59–60) own admission, there is no experimental evidence that could support the existence of the scaffolding once it has been removed (Fry 1999, pp. 185–187).

Whilst Cairns-Smith's sketches are undoubtedly an important step towards hybrid theories, the first theory to explicitly discuss the origin of life as two separate-but-necessary events was from Freeman Dyson (1984; but see also this implication in Calvin 1969). Unfortunately, Dyson misunderstood Cairns-Smith as presenting a replicator-focused theory where clay replicators are supplanted by enzymes, cells, and then genes—rather than seeing this as an accumulative process for replicators and replicators-in-organisms. Dyson instead frames his ideas as relating to the dual origin of the chemistry of metabolism and replication. Dyson claimed to be heavily influenced by Oparin's (1924) reasoning about the essentiality of the metabolic cell, which could form by something like colloid chemistry alongside related catalysts that he thought were likely to be protein-based (citing Fox, 1980). Dyson also claimed influence from

Manfred Eigen (who in turn owed much to Haldane 1929) that the first replicators, which he thought were likely to be RNA-based, were parasites of the first cells. Although this has similarities to a takeover event, taking inspiration from symbiotic theory (Margulis 1970), Dyson suggests that these parasites eventually evolve to become symbionts that take on the specialised role of storing hereditary information. This has been influential, with Addy Pross (2012, pp. 158–159) suggesting that life began with the symbiosis of a non-metabolically active replicator and a metabolically active cell. But, much like for Cairns-Smith, it is still unclear how any of this process of the origin of parasites and their specialisation might occur; moreover, in the present day, it is much more widely accepted that parasites are not fixed on an evolutionary trajectory to evolve to become more benign (i.e. into symbionts), but rather face a trade-off between virulence and transmission that may be balanced in multiple ways in different ecologies (Anderson and May 1991, pp. 648–652).

Hybrid theories like these are important for the understanding of organism- and replicator-first theories because they suggest that it is possible to assimilate their insights into a coherent synthesis. Indeed, through the way that metabolism-first theories are really organism-first, their integration is seemingly an extension of a broader duality of focus in evolutionary biology. Dawkins (1982, p. 6) used the optical illusion of the Necker cube to illustrate how there are multiple compatible ways of approaching the same truths in evolutionary theory: focusing on genes or organisms (although, it must be admitted that Dawkins (1982) advocates the gene's-eye view because he thinks it is less likely to lead to error). So too it seems with the origin of life: an organism-first approach focuses on how life takes energy from the environment to provide the fuel for reproduction, whilst a replicator-first approach focuses on what the fuel is used to do (i.e. the more making that leads to evolution by natural selection). Of course, different theories can still make conflicting statements, especially about the order of events (Fry 1999, pp. 150–178): organism-first approaches tend to assume that protein-based metabolism came first and replicators much later as a derived form of information storage, whilst replicator-first approaches tend to assume that RNA-based replication came first and cellular metabolism much later as a derived form of increasing replicability. These disagreements are not trivial, but both sides seem to willingly foster a false dilemma.

Therefore, despite the progress that has been made, it seems indisputable that the key challenge for understanding the origin of life is not the mastery of the experiments or hypotheses that have been proposed, but rather an issue of imagination, because explanations still start from some kind of improbable soup and eventually run up against an insurmountable problem. For organism-first theories, it is not clear self-organising protein sets are capable of gradual evolution in reproducing errors. Consequently, natural selection is too weak, and the content of the earliest cells is likely to evolve more by random drift than by anything else. For replicator-first theories, it is not clear that simple replicators are capable of increasing complexity because natural selection is so strong that they are selected for their replicability over their functionality. For hybrid theories, some aspects of complexity may seem to be able to evolve more gradually (e.g. by takeover), but both problems from organism- and replicator-first theories still persist. The common feature that unites the problems for the different theories is tangling with the wrong kind of natural selection at the claimed origin of life.

* * *

It would appear that theory has no unambiguous answers as to whether life elsewhere in the universe would exhibit maladaptation because there is still much that is unknown in the theory of the origin of life. But the core scientific insight of maladaptation may provide some further reasoning. The persistent notion that life must arise from an improbable soup is misleading because it supposes that, once it has miraculously arisen, natural selection can be expected to drive the increasing efficiency, diversity, and complexity of life. With the possibility of maladaptation, natural selection can also drive extinction. Maladaptation provides an apt description of what is seen in organism- and replicator-first theories of life when chemical evolution by natural selection favours those entities for their reproducibility or replicability at the expense of their functionality, which may ultimately lead to their extinction. By contrast, biological evolution often leads to adaptation, which provides life with many possible futures. Therefore, it is as if there is an overriding tendency to favour maladaptation that separates closed-ended chemical evolution from open-ended biological evolution. How can this be overcome?

Given that biological evolution has its origins in chemical evolution, there must be some way that the evolutionary trajectory of chemical evolution can move in a different direction. There are two basic solutions at opposite extremes, which have implications for the connection between life and maladaptation. First, a chemical system could be constrained so that natural selection is inherently predisposed to favour adaptation, whereupon maladaptation would be a derived property of earthly life that may not be shared by life elsewhere in the universe. This would seemingly require the miraculous origin of a biological system that is fully assembled despite the maladaptive tendency of chemical evolution, which is totally unlike those simple chemical systems that have been discussed by replicator- and organism-first theories of the origin of life. Second, a chemical system could be inherently predisposed towards maladaptive extinction, which would make maladaptation a more general property of all life. Such a chemical system would only lead to something like life through miraculously good fortune, by randomly walking sufficiently far down the path of successive adaptations to establish a sufficient buffer against maladaptive extinction. Of course, the real solution might fall somewhere between these two extremes, but unfortunately neither basic solution is particularly compelling. Being predisposed towards maladaptation fits with the identified pattern of chemical evolution, and the current state of life on Earth seems like a farfetched outcome by chance alone. The origin of life is a perplexing problem, demanding imagination.

Consider a thought experiment concerning puddles. The puddles are temporary entities that are produced by rain falling onto any slight depression on a surface, like a rock-flat. The puddles may provide the opportunity for enhanced erosion as, whilst it lasts before evaporation, the liquid may erode the rocky surface. Some puddles may be wider, which gives them a larger surface area for evaporation. Deeper puddles with a smaller surface area for evaporation are more likely to persist for longer periods of time. As a result, deeper puddles are also likely to erode into the rocky surface faster. With repeated cycles of rainfall and evaporation, those puddles that started out deeper may enable greater erosion into the rocky surface. As they do, puddles may horizontally erode the walls in between them leading to mergers. Alternatively, puddles may encounter harder seams of rock that create barriers that block mergers or (with vertical erosion) divide a puddle in two. Despite such constraints, puddles that start out deep would erode to form wider basin-like shapes. Although this may give them a wider watershed, this is more likely to be disastrous for the persistence of the puddle over time, as it affords it a larger surface area for evaporation. Over long periods of time, puddles may become even more temporary.

The expected evolutionary trajectory of puddles may be interrupted by a takeover event, whereby the traits of puddles begin to be increasingly shaped by their dusts rather than their starting points (cf. Cairns-Smith 1982). As well as puddles, dust may also impact the evolution of a rock-flat. Dust from nearby or far away may blow into puddles on the wind, becoming trapped by the water body (cf. Oparin 1967). The liquid of the puddle may prevent the dust from being blown away. It is conceivable that different dusts may alter the erosion process, for example by silting up the puddles to decrease erosion or by increasing acidity to increase erosion. Deeper puddles are more likely to persist for longer periods of time, which may also give them a higher chance of gathering more dust over time. In a puddle, the dusts that are likely to persist might be supposed to be those that are chemically inert or stable, but such dusts would be more likely to silt up puddles than anything else, which would dry a puddle out faster. Instead of inert chemistries (*contra* Dawkins 1976, p. 12), dusts that become more numerous in the puddles are likely to be those that are produced from erosion by the puddles. Indeed, those dusts that can increase erosion to produce more of themselves are likely to become more

numerous. Presumably, it would be remarkable if dusts were highly effective at increasing erosion. But any increasing erosion only hastens disaster
for the puddle, as its larger surface area leads to faster evaporation—and
the dust blows away as the puddle dries out.

Independently, puddles and dusts could be described as changing
through a gradual evolutionary process. Indeed, in each case, it is as if
there is natural selection to favour faster erosion, leading onto faster
evaporation. Further, puddles and dusts have variation, heritability, and
reproduction, which have been identified as the criteria for natural selection (Darwin 1859; popularised in this form by Lewontin 1970). Yet
the evolutionary process is very unlike biological evolution by natural selection, not least because the change leads the puddles and dusts
towards self-destruction: the puddles dry out faster, and the dusts are
not produced when the puddles are dry. In this regard, the proposed criteria for natural selection appear vastly too lax in describing a kind of
chemical evolution by natural selection that is an evolutionary dead-end
(which has been neglected by its proponents; e.g. Godfrey-Smith, 2009,
pp. 135–145).

Alternatively, together, puddles and dusts can form a simple protobiological system of dependent chemical evolution. Puddles provide natural
individuation like cellular compartments, defining separate protoorganisms as hubs of the metabolic activity of erosion. A puddle has limited
chemical properties on its own, as a water body that is formed by rainfall,
but may come to have extrinsic properties of shape and watershed that
are determined by the historical patterns of erosion at its place in space.
Puddles have their own evolutionary trajectory based on their starting
conditions (i.e. on the variation on the rock-flat), which tend towards
their self-destruction. When dust arrives in the system, it takes over the
driving of evolutionary change. Dust naturally provides a puddle an
inheritance through intrinsic properties, which would be taken with the
puddle if it was transplanted to another location on the rocky surface. As
such, the presence of dust arrests some control over the existence of the
puddle from the environment (cf. Huxley 1942, pp. 515–516). Initially,
the dusts blow in from the wider environment, and so the inheritance

they provide arises by chance (cf. Dyson 1982). Over time, the dust may come to be dominated by simplistic protoreplicators, with more numerous dusts being those that are produced by the metabolic activity of erosion—and, of those, the dusts that can slightly increase erosion to produce more of themselves. Whilst driving the evolution of puddles, the presence of such dusts does not by itself alter the direction of evolutionary change, as they silt up puddles to still hasten their destruction, which is ultimately an act of self-destruction as the process of erosion ends the pathway to replication.

Whilst the chemical evolution of both puddles and dusts is independently predisposed towards self-destruction, the chemical evolution of puddles through dusts may not be. Whilst dusts silting up puddles may be the expected trajectory, dusts could have other effects, like floating to the surface of puddles to reduce the rate of evaporation (e.g. by reflecting sunlight). In ways such as this, it is possible for chemical evolution to favour dusts that make a functional contribution to puddles in a way that does not rapidly lead to their self-destruction, which is much more like the open-ended pattern of biological evolution by natural selection.

In simple terms, it is possible to recognise two distinct selective pressures on the system of puddles and dusts, which are neglected for all current theories of the origin of life (including Cairns-Smith 1982). There is a selective pressure for maladaptation when puddles and dusts are favoured as independent chemical systems. The predominant replicator-first theories do not recognise that organisms are independently evolving entities, and so they assume that replicators can exploit them without damaging the hospitable environment of the puddles that they need to replicate. There is also a selective pressure for adaptation when puddles and dusts are favoured as dependent chemical systems. The predominant organism-first theories do not recognise that replicators can act as independently evolving entities, and so they assume that organisms can evolve to pass on traits that are functional (rather than just replicable). The establishment of an open-ended pattern of biological evolution by natural selection really stems from the coevolution of dependency among entities that act as replicators and organisms.

If a chemical system starts with the natural selection of organisms and replicators in complete independency, then it is likely to go extinct. For example, if a gene-like replicator did miraculously arise in a puddle, as an inert autocatalyst that could exploit the products from the erosion to make more copies of itself (cf. Haldane 1929), it would undergo rapid chemical evolution. By its extraordinary good fortune to appear in a hospitable environment, the replicator is likely to drastically increase in frequency within the puddle. Small replication errors would create mutant variants, enabling natural selection to increase the frequency of some replicators faster than others. Indeed, those replicators that are faster at exploiting the products from erosion would be favoured by natural selection (cf. Spiegelman 1967). There is no way out for natural selection to favour replicators in any other way than as a harmful parasite that eventually drives its own extinction in the puddle (*contra* Dyson 1984). For instance, the puddle is likely to fill up with waste products that make the chemical environment increasingly inhospitable to the replicator. Therefore, if a replicator did arise at this stage, it would be likely to drive itself to extinction by destroying the environment that it needs to thrive.

Dependency can prevent extinction through natural selection favouring replicators for their contributions to the survival and reproduction of organisms. It is reasonably likely that the natural selection of protoorganisms and protoreplicators starts with them in a close dependency. This can be seen with puddles and dusts, where their simple origins mean that they start out as highly ineffective organisms and replicators—and natural selection on their independent traits may only serve to make them more ineffective. So puddles are very temporary, and dusts do not persist long enough to replicate themselves. Consequently, if a dust is to stand any chance of replicating, it is likely to need to extend the lifespan of the puddle that it is within in order to do so. Indeed, it may take many wet–dry cycles for a dust to bias chemical change to produce just one extra molecule of itself. Therefore, successful dusts may not be those that are the fastest replicators within a puddle, but rather those that improve the hospitableness of the environment of puddles in general (because they may blow away before they replicate; cf. Wright 1931), which may eventually enable them to spread faster through a population of puddles.

The flaw in the reasoning behind the miraculous origin of the replicator in an improbable soup now becomes more obvious. The miraculous replicator is a 'higher technology' (*sensu* Cairns-Smith 1985, p. 100) of modern life that would be incompatible with the fragile protoorganisms that were around at life's origin. It is easy to overlook that the large, multicellular modern-day organisms that attract our attention work in obviously derived ways, including the capacity for their replicator to make copies of itself faster than an organismal generation time. This extraordinary state of affairs is only possible because of a single-cell bottleneck in the germline and the high-fidelity replication of genes into somatic cells, allowing replicators to pass instructions through a reliable dissemination; albeit that the seemingly inevitable evolution of cancerous mutations may lead to the eventual extinction of the hospitable environment of the multicellular organism.

A more likely origin of life involves the probable soup that arises in systems with natural individuation and inheritance, as protoorganisms and protoreplicators gradually coevolve. If protoreplicators start to replicate much faster than protoorganisms, they are liable to end up producing maladaptations that can threaten the continuation of their coevolution. Similar things happen with social learning, where traits are copied from one individual to another. As a secondary, non-genetic system of inheritance, natural selection may favour traits on the basis of their replicability rather than their functional contribution to individual fitness (see e.g. Boyd and Richerson 1985, pp. 7–11). The only conceivable way that would reliably lead to adaptations (whether through learning rules or learnt behaviour itself) spreading through a population is when the non-genetic replicator is favoured or disfavoured because of its effects on individual fitness. This requires a lasting association between the non-genetic replicator and its fitness effects on individuals, which would arise when replication occurs more slowly than an organism's generation time (most likely as constrained by low-fecundity or -fidelity replication). Consequently, natural selection primarily acts on traditions or cultures that arise from non-genetic replicators in groups of individuals—and much of the learnt behaviour (that is idiosyncratic or 'fashionable' to individuals) within each group may be maladaptive (see also Boyd and Richerson 1985, pp. 282–289).

Linking the ancient origin of life to the modern life forms all around us, the consideration of a potential new takeover event by non-genetic replicators in the history of life on Earth provides a penetrating—but nuanced—insight into the relationship between maladaptation and life (in all its forms). A maladaptive trait is favoured by natural selection because it increases the relative fitness of an allele through an expense to the absolute fitness (or survival and reproduction) of the individual organism with the allele. A maladaptive trait requires an interaction among entities with different alleles, and it is through the interaction that the allele reaps its relative advantage. Consequently, the allele for maladaptation finds a way of increasing its relative fitness through the trait *in sensu stricto*—in that its benefit to that allele can be truly instantaneous by destroying what currently is; by contrast, an allele for adaptation only realises the benefit of that adaptation with the eventual production of more surviving offspring—by creating what will be. This way of conceptualising maladaptation suggests that all maladaptations could be conceived as involving a replicator replicating relatively faster than the organism it resides in absolutely produces organisms.

With this in mind, the question at the start of the chapter can be returned to: should all life exhibit maladaptation? The question has been addressed by focusing on the origin of life because it is likely to be a similar problem anywhere in the universe, whereas the familiar features that we might imagine are shared with all life are difficult to disentangle from those features of earthly life that are common to all but entirely contingent on the direction that evolution has taken. Without needing to identify a hard boundary between life and nonlife, close to the origin of life, chemical evolution has an identified tendency that favours maladaptation, as replicability trumps functionality, which may lead evolving chemical systems to extinction. The origin of biological evolution by natural selection is a great domestication event where those earliest protoreplicators coevolve to serve the interests that they have in closer alignment with their protoorganisms. It is plausible that this may largely arise because of the constraints on the protobiological system of chemical evolution that favour adaptation (rather than e.g. chance generating successive mutations in that direction). Whatever the origins, the future of the relationship may well be fraught with adaptation and maladaptation because, even going all the way back to life's origins, there do not appear to be any mechanisms that prevent maladaptation from occurring: it is an ever-present possibility from natural selection. Indeed, alongside cumulative adaptation, natural selection can lead to cumulative maladaptation as well. So, inasmuch as theoretical speculation can provide evidence, maladaptation appears to have been present right from the very beginnings of biological evolution by natural selection—as part of the foundational architecture of life. By implication, all life anywhere in the universe should be expected to be subject to maladaptation, which is the inheritance of biological evolution from its origins in chemical evolution, as the visible proof of an unresolved flaw in the design of all living things. Such a conclusion starts to have meanings that transcend its scientific basis, which must now reluctantly be addressed in the final chapter.

8

Revisiting the design argument

Now, at the last, those wider implications of maladaptation are considered, despite some reticence. It is almost a tradition in contemporary books about evolutionary biology to have a final chapter about humans, which tends to overstretch the otherwise scientific discussion to support some banal philosophy, or to otherwise make a series of crude generalisations about human nature. Despite the poor precedent, having reached this point, it is obvious that there is a danger in not engaging with the topic because, as David Hume (1739, p. xix) remarked: ''Tis evident, that all the sciences have a relation, greater or less, to human nature; and that, however any of them may seem to run from it, they still return back by one passage or another'. Unfortunately, at present, there are too few sufficiently strong examples to provide a review of maladaptation in humans. But the general consideration of maladaptation, in its way, does 'throw light' on a received understanding about our place in the universe.

At the very beginning of the book, it was asked: why do living things appear designed, and what are they designed for? Evolution by natural selection is now well accepted to be the scientific mechanism behind design in nature, but the second part of the question about the purpose of design is more contentious because it treads on the toes of philosophers and theologians. Over Darwin's lifetime, his own views on the matter hardened such that, by 1876 when he was writing his *Autobiography*, Darwin (1958, p. 87) wrote:

> The old argument of design in nature, as given by Paley, which formerly seemed to me so conclusive, fails, now that the law of natural selection has been discovered. We can no longer argue that, for instance, the beautiful hinge of a bivalve shell must have been made by an intelligent being, like the hinge of a door by man. There seems to be no more design in the variability of organic beings and in the action of natural selection, than in the course which the wind blows.

So, for Darwin, the need for there to be anything more to say about design in nature vanished with the scientific discovery of evolution by natural selection because whatever appearance of design there is arises from a natural process. The histories of evolutionary biology have always been sympathetic to this self-stated achievement, agreeing that 'Darwin firmly drove the idea of God out of nature' (Browne 1995, p. 543). Consequently, it is now mainstream presentation of historical fact that 'If Paley clearly formulated the problem of adaptation, then Darwin . . . decisively solved it' (Gardner 2009, p. 861). Such reasoning underlies Dawkins's (2006, p. 103) case in *The God Delusion* that 'There has probably never been a more devastating rout of popular belief by clever reasoning than Charles Darwin's destruction of the argument from design', as the science of evolutionary biology is presented in support of atheism.

The domination of this atheistic narrative in describing the foundations of evolutionary biology has not gone wholly unchallenged. Gould (1999, p. 4) famously put forward his case for 'nonoverlapping magisteria' in *Rocks of Ages*, whereby

> Science tries to document the factual character of the natural world, and to develop theories that coordinate and explain these facts. Religion, on the other hand, operates in the equally important, but utterly different, realm of human purposes, meanings, and values—subjects that the factual domain of science might illuminate, but can never resolve.

Gould is not trying to argue that the claims of science and religion have not come into conflict over their history, nor is he trying to be prescriptive about where the boundaries between science and religion might lie. Instead, Gould (1999, pp. 69–70) is acknowledging that there has often been a successful working relationship between science and religion 'established by long struggle among people of goodwill in both magisteria'. But whilst Gould (1999, p. 192) firmly states that 'Darwin did not use evolution to promote atheism, or to maintain that no concept of God could ever be squared with the structure of nature', elsewhere he agrees that Darwin made a 'radical argument against Paley' that 'inverts Paley's world' (Gould 2002, p. 121).

At least, then, everyone seems to agree that Paley's design argument is untenable. In this regard, it would seem that Paley has become the resident villain of evolutionary biology who is only discussed as a straw man—as the person who got the explanation of design wrong. Even Christian apologists like Alister McGrath (2008, p. 102) have suggested that 'Paley's approach to natural theology has to be seen as an intricately and beautifully constructed house of cards'. With such differing perspectives finding consensus here, it would seem that the tide of history has long been turned against Paley. Having returned to reconsider his work to understand the foundations of evolutionary biology, the consensus seems prodigiously wrong. At this point where Paley's case has seemingly been so decisively lost, perhaps the time has now come for a turning of the tide.

In short, the source of the disagreement with the consensus is about Paley's question. It seems that the narrative that the discovery of evolution by natural selection destroyed Paley's design argument makes a category mistake (*sensu* Ryle 1949), whereby the scientific question of design that Darwin answered does not help to address the theological (or philosophical) question of design that Paley posed. This is not meant as a 'tedious cliché . . . that science concerns itself with *how* questions, but only theology is equipped to answer *why* questions' (Dawkins 2006, p. 80); unlike Gould's nonoverlapping magisteria that seems to relegate religion to the pursuit of subjective truth, it is clear that science and theology are both concerned with objective facts. Instead, there are reasonable grounds for thinking that Paley, with his goals in mind, would have defended his design argument today for reasons that still make sense after the discovery of evolution by natural selection.

Before exploring the evidence of this, in the present day and age, there is a need to restate what the theological question of design is all about. In a secular age, it might seem like the need for natural theology has been completely undermined. But this stems from a misunderstanding of its goals, which for Paley had little to do with the existence of God per se. It can be confusing when reading Paley (1802, p. 415) because he sometimes states that he is out 'to prove the existence of an intelligent Creator', but the emphasis here rests on the character rather than the existence of God, which is clear from the content of *Natural Theology*. Moreover, Paley is not really talking about God with all the popular implications of the term. Indeed, the tradition of natural theology that Paley follows has its roots in Ancient philosophers who had a range of beliefs about the existence (or not) of the Greek pantheon of gods, but they still debated the character of what Paley called God. Put more agnostically, the theological question of design is about the character of the creative forces that are behind everything. Those creative forces underpin the relationships we have with ourselves, each other, and nature. This includes how we perceive the world around us in the senses that we have, the way that we feel (and think) and the shape of our physical bodies. Through intersecting every aspect of our lives, Paley (1802, p. 542) stated: 'The existence and character of the Deity, is, in every view, the most interesting of all human speculations'. Whatever is believed about the conceptualisation of those creative forces, it is a fair question to ask, as Paley did: have we been created in our own interests? Because, whether we like it or not, 'Our happiness, our existence, is in [their] hands' (Paley 1802, p. 541).

To understand Paley's response to this question, it is first necessary to reconcile his positioning of the creative forces as God. For the philosophers and theologians at the time, it was scarcely meaningful to debate the existence of God because it was largely taken to be self-evident, which is often much misunderstood by modern audiences. Most famously in *Summa Theologica*, Aquinas grouped together five arguments that 'proved' (in an archaic sense) the existence of God, and he attached particular weight to the fifth argument from first cause (see Aquinas 1998, pp. 243–256). In brief, building on Aristotle, Aquinas claimed that the world is governed by chains of cause and effect, and there must have been a first act of creation from which all subsequent acts of creation were effects; he described this first cause as God. It sometimes needs clarifying for modern audiences that this has nothing to do with 'what came before the big bang' through which the universe came into existence. In refutation of a similar misunderstanding to a different historical audience of the fourth century CE in *City of God*, Augustine repeatedly argues against the claim that the pantheon of Roman gods included the God that deserves one's attention because, in Greco-Roman mythology, the gods were created by other primordial deities (and so on, backwards in time leading to an infinite regress of causal responsibility; see Augustine 2003, pp. 291–292). Paley (1802, p. 412), demonstrating his limited definition of God as a first cause in *Natural Theology*, makes a similar case to Augustine when he argues 'that whatever the Deity be, neither the universe, nor any part of it which we see, can be He' because that 'supposition involves all the absurdity of self-creation, i.e. of acting without existing'. God as first cause bypasses such absurdity by tautology, whereby whatever the unknowable first cause that brought causality into existence is becomes the minimal (but in many ways essential) notion of God by being the creator of the world. Other gods are idolatrous pretenders.

As was recognised with diligence by Aquinas, and later famously emphasised with delight by Hume (1779) in *Dialogues Concerning Natural Religion*, such a tautological argument for the existence of God as a creator does not infer their character. For Paley as much as any Christian, the character of God is most persuasively established in the revelation of the Bible (Paley 1802, p. vii; see also Paley 1794). But, where such revelation is doubted, Paley accepted a need to establish the character of God from nature. In *Natural Theology*, Paley (1802) set out to demonstrate the most essential characteristics of the Abrahamic God using observations from nature. A single characteristic is critical in permeating through all others: benevolence, which refers to a general well-meaning. As a result, above all else, Paley sought to find evidence for the benevolence of God in the design of individual organisms, using observations about the complexity, regularity and purposiveness of their traits.

In simple terms, there are but three basic ways the evidence could point. Paley (1802, pp. 465–467), quoting from his *Moral Philosophy* (1785, pp. 39–40) talks through each in turn:

> When God created the human species, either he wished their happiness, or he wished their misery, or he was indifferent and unconcerned about either. If he had wished our misery, he might have made sure of his purpose, by forming our senses to be so many sores and pains to us, as they are now instruments of gratification and enjoyment: or by placing us amidst objects, so ill suited to our perceptions as to have continually offended us, instead of ministering to our refreshment and delight. . . . If he had been indifferent about our happiness or misery, we must impute to our good fortune (as all design by this supposition is excluded) both the capacity of our senses to receive pleasure, and the supply of external objects fitted to produce it. But either of these, and still more both of them, being too much to be attributed to accident, nothing remains but the first supposition, that God, when he created the human species, wished their happiness; and made for them the provision which he has made, with that view and for that purpose. The same argument may be proposed in different terms; thus: Contrivance proves design: and the predominant tendency of the contrivance indicates the disposition of the designer.

So Paley was persuaded that the evidence suggests the creative forces acted in our interests, giving us bodies and minds that are largely designed to help us navigate through life.

Everyone seems to agree that 'Paley's observations are undoubtedly correct' (Gould 2002, p. 120), but disagreements emerge over their meaning. Instead of design arising from benevolence, evolutionary biologists have overwhelmingly thought, against Paley's (1802, p. 466) argument, that whatever happiness is possible for us arises from 'the incidental effect of organisms struggling for their own benefit' (Gould 2002, p. 121). In evolutionary theory, there is no guarantee that natural selection gives us traits that make life worth living. For even the moderate voice of Gould, this is not a matter of philosophical interpretation, but rather is a matter of scientific fact; evolution by natural selection does not generate 'any larger harmony that might embody God's benevolent intent'

(Gould 2002, p. 127), which simply lies beyond the justifications of the science.

Natural Theology contains many arguments that Paley uses to defend his position against critics. Before assessing his pre-emptive response to the scientific discovery of evolution by natural selection, it is worth establishing that Paley was not a Leibnizian optimist, thinking that this world is 'the best of all possible worlds'. Whilst, for example, his contemporary Robert Malthus (1798, p. 162), in a seemingly rare moment of optimism, made the extraordinary claim that 'there is no more evil in the world than what is absolutely necessary', Paley never entertained such speculation. Paley held an empirical philosophy that invites scepticism when making claims that lie beyond what can be reasoned from evidence. Further, despite advocating an early utilitarian ethic that sometimes used the language of the moral calculation of happiness in passing (e.g. Paley 1794, pp. 148–149), Paley (1794, pp. 149) never used the metaphor of calculation to displace 'the exercise of wisdom, judgement, and prudence', unlike later utilitarians who entertained speculative moral calculations (e.g. Mill 1863). With a grounding in empirical philosophy and utilitarian ethics, Paley (1794, pp. 252–253) explicitly framed his outlook against the theodicy that this is the 'best of all possible worlds' when he argued that 'We cannot judge of [Leibnizian] optimism, because it necessarily implies a comparison of that which is tried with that which is not tried, of consequences which we see with others which we imagine, and concerning many of which, it is more than probable, we know nothing'.

Equally, Paley was not a Voltairean pessimist. In *Candide*, Voltaire (1759) satirised the Leibnizian optimism of the naïve Professor Pangloss, who always found a way to shut his eyes to the evil he was seeing on the basis of his blind faith in this being the best of all possible worlds. Detractors of natural theology like Voltaire focused on whether specific evils were strictly necessary, like famously the 1755 Lisbon earthquake, which led critics to dub their outlook as being pessimistic (though it is unlikely that Voltaire would have accepted this label). Again, given his philosophical and ethical outlook, Paley was not sympathetic to such cherry-picking of specific evils in the face of their complex underpinnings. With respect to the general laws of nature, Paley (1802, p. 493) argued that 'we are uninformed of their value or use; uninformed consequently, when, and how far, they may or may not be suspended, or their effects turned aside, by a presiding and benevolent will, without incurring greater evils than those which would be avoided'. An argument from ignorance is dangerous because it can cut both ways, so Paley had to go a step further, returning to his core point. Whilst Paley (1802, p. 56) freely admits that 'imperfection, inaccuracy, liability to disorder, occasional irregularities, may subsist in a considerable degree', Paley (1802, p. 467, quoting Paley 1785, pp. 40–41) rested his argument on the observation that wherever such 'Evil, no doubt, exists . . . [it] is never, that we can perceive, the object of contrivance'. Consequently, through knowing the goal of design in nature, ignorance of the cause of imperfections should not sway the argument (see Paley 1802, p. 8).

Paley positioned his outlook as one of moderation. Paley did not feel he could demonstrate the maximal good of optimism from the evidence of nature, nor could he find the grounds for overriding evil in pessimism. Instead, from the evidence of nature, he saw benevolence, which is in many ways a minimal kind of goodness. In explicit contrast to Leibnizian optimism, Paley (1794, p. 252) put forward that 'We can judge of beneficence, because it depends upon effects which we experience, and upon the relation between the means which we see acting and the ends which we see produced'. Of those ends, Paley (1802, p. 463) summarised that 'happiness is the rule, misery the exception'. In this way, relying on the strength of his interpretation of traits as adaptations, Paley pursued a simpler premise for his design argument than many of his would-be critics rely upon to dismiss the whole enterprise of natural theology.

More as a point of clarification than defence (because there was no equivalent explanation at the time), Paley seemingly pre-empts something like the scientific discovery of evolution by natural selection being used to challenge his argument. Paley's God is the author of design in nature by being the first cause that set in motion the causality in nature. Paley (1802, pp. 419–420) also discusses the possibility that 'There may be many second causes, and many courses of second causes, one behind the other, between what we observe of nature, and the Deity'. Whilst Paley did not know of evolution by natural selection, there is nothing inconsistent with importing it *ex post facto* as a theory of secondary causation. Paley (1802, p. 420) suggests that design 'may be the result of trains of mechanical dispositions' but insists that they are 'fixed beforehand by an intelligent appointment, and kept in action by a power at the centre'. So, for Paley, there would always be a need to look beyond nature to the first cause to answer his theological question of design. It should therefore be obvious that Paley was not presenting God as the scientific explanation of design, as if God were its secondary cause like a human craftsman (*contra* McGrath 2008, pp. 97–99; see later in this chapter).

Consequently, Paley's design argument appears robust to the discovery of evolution by natural selection. The steps of his core reasoning would appear to be very capable of flexing to maintain its logical coherence, which is substantially because, unlike other contemporary treatments of natural theology (e.g. Malthus 1798), Paley does not make many claims. Paley's argument establishes a few key points and then rigorously defends them against potential criticisms, which in one way or another all seem to rest on the identification of the goal of design in nature. Consequently, it is difficult to see why the discovery of evolution by natural selection should entail the destruction of Paley's design argument—especially when Darwin preserved its logical structure (e.g. Darwin 1859, p. 211). Here instead, there is a need to recognise that, when Darwin co-opted Paley's reasoning for his presentation of natural selection, he not only switched the external cause of design from God to natural selection (in a way that preserved the benevolence of the creative forces), but in doing so he also switched from a theological to a scientific question of design.

Accepting this switch happened, it is a further point to demonstrate that others also recognised the switch of questions at the time. Perhaps one of the single clearest demonstrations of this came during the infamous Huxley–Wilberforce debate in 1860. Many opinions were expressed on the day, including those who staunchly represented the Christian establishment. From this contingent, Frederick Temple, who later went on to become Archbishop of Canterbury, publicly voiced the Anglican establishment's response to Darwin's ideas, which was soon to be the widely accepted Christian response: evolution by natural selection is simply God's method of design in creation (McGrath 2011, p. 164). The perspective was hardly a boldly original take, not least because it may as well have been lifted straight from Paley's (or indeed Augustine's or Aquinas's) discussion of secondary causes.

Many arguments that treat Paley as a 'straw villain' seem unaware of the details of his design argument. Interestingly, despite being largely overlooked, arguments that more seriously set out to attack Paley's design argument seem to require an extra step of reasoning beyond merely asserting its incoherence after the scientific discovery of evolution by natural selection. The case against Paley rests on a more subjective, emotional argument about the perceived nastiness of nature. Darwin (1887c, pp. 311–312) led the way when he famously wrote: 'I cannot persuade myself that a beneficent and omnipotent God would have designedly created the Ichneumonidae with the express intention of their feeding within the living bodies of Caterpillars, or that a cat should play with mice'. Feeding into the Victorian zeitgeist, Darwin's pessimism spoke to Tennyson's view of nature as 'red in tooth and claw', where there is a ruthlessly free competition in the struggle for existence, like that older metaphor of the Hobbesian war of all-against-all that leads to immense suffering. In a later generation, Huxley (1942, pp. 439–440) expounded: 'Natural selection, in fact, though like the mills of God in grinding slowly and grinding small, has few other attributes that a civilized religion would call Divine'. Similarly, Haldane (1968, p. 45) concluded that 'If then animals were designed, they were designed for mutual destruction'. And even Gould (2002, p. 121; see also p. 127), despite his nonoverlapping magisteria, admits: 'What a bitter cup Darwin offers us, compared to Paley's sweet promises'.

However it is stated, the case against Paley's design argument from the nastiness of nature reduces to the problem of evil (or, otherwise, pain or suffering), which is probably the longest-standing and most discussed reason for doubting the existence of a benevolent God. Darwin (1887c,

p. 311) explicitly acknowledged this, writing that, for him, 'This very old argument from the existence of suffering against the existence of an intelligent First Cause seems to me a strong one'. Although the examples from nature that Darwin and others invoke differ from mainstream presentations on account of their knowledge of natural history, evolutionary biologists have added shockingly little to this debate. In short, this is because no one was in any doubt before Darwin that life was a struggle that sometimes involved the most unbearable suffering—least of all Christians who believed that almighty God was tortured and killed in the person of Jesus Christ, who is even reported to have cried out for its end during his crucifixion, 'My God, my God, why have you forsaken me?' (Matthew 27: 45–56, NIV).

There are a diversity of Christian responses to the problem of evil that have accumulated over the years. For Paley, suffering is not for any notion of spiritual development in the Irenaean tradition, which justifies the benevolence of God, in the pitiless words of Haldane (1985, p. 97), 'based on the celebrated hypothesis that two blacks make a white'. Instead, Paley (1802, pp. 493–494, all emphases original) admits that whilst 'no universal solution has been discovered . . .. The most comprehensive is that which arises from the consideration of *general rules*' whereby 'cases of apparent evil, which *we* can suggest no particular reason, are governed by reasons, which are more general, which lie deeper in the order of secondary causes'. So Paley casts doubt on instances of, perhaps especially, apparent 'natural evil' through appeal to the necessary constraints of natural order. One failing here is that Paley does not use some known examples to show how a 'useful' general rule could lead to apparent evil. Instead, he rather weakly presents this as a precautionary principle, which is difficult to unreservedly maintain.

Paley seems aware that this is not a wholly satisfactory treatment of the issue. Consequently, for more tangible examples of 'moral evil', Paley (1802, p. 511) elaborates further by drawing upon the more popular Augustinian tradition:

> The mischiefs of which mankind are the occasion to one another, by their private wickednesses and cruelties; by tyrannical exercises of power; by

rebellions against just authority; by wars; by national jealousies . . . or by other instances of misconduct either in individuals or societies, are all to be resolved into the character of man as a *free agent*. Free agency in its very essence contains liability to abuse. Yet, if you deprive man of his free agency, you subvert his nature. You may have order from him and regularity, as you may from the tides or the trade-winds, but you put an end to his moral character, to virtue, to merit, to accountableness, to the use indeed of reason.

Paley (1802, p. 512) ultimately ties this back to his overarching explanation of general rules of secondary causes by considering how 'passions are strong and general' and so liable to apparent evil from excess in much the same way as before. So, for Paley, these more obvious examples of genuine evil, which seem to be exclusively moral evils, are the responsibility of individuals from their use of free will, rather than God their creator who gave them their free will. The existence of evil is then the inexorable price of agency, which is taken to be an essential facet of human nature.

Such reasoning may help to reconcile the benevolence of God with the existence of evil, but, as arguments appeal to the nastiness of nature, it is another step to respond to the emotional reaction. Paley, as much as Darwin, might have been revolted by the life cycle of an Ichneumon wasp if he were aware of it. As close as he gets, Paley (1802, p. 481) remarks that 'animals *devouring* one another, forms the chief, if not the only instance, in the works of the Deity, of an economy, stamped by marks of design, in which the character of utility can be called in question', but concludes that 'it is probable that many more reasons belong, than those of which we are in possession'. So again Paley freely admits his doubt in being able to recognise the goal of such design. However, Paley does not see this as damaging to his argument, not least because predation is an 'ecological' phenomenon, and Paley founds the strength of his argument on observations about the utility of individuals' traits to those individuals (rather than to other individuals in the economy of nature). Consequently, Paley (1802, p. 482) simply reasserts that 'in a vast plurality of instances, in which *contrivance* is perceived, the design of the contrivance is beneficial'. Therefore, Paley would protest the nastiness of nature because, when he looks on the purpose of its design, happiness is still the rule and misery the exception.

Whilst many evolutionary biologists might not agree with Paley's design argument, there is little justification for the claim that Darwin somehow demolished Paley's case. On this account, Paley deserves to be remembered for more than being the straw villain who got the scientific explanation of design in nature wrong, which seems to stem from a lazy deference to later interpreters—not least Darwin—rather than a serious attempt to understand Paley in his own terms. Perhaps Paley, along with many other forerunners in the prehistory of evolutionary biology, has suffered such characterisation because of a tendency to view his beliefs as unintelligible. There is a famous quote, popularised on the first page of Dawkins's (1976) *The Selfish Gene* from George Simpson (1966, p. 472), that 'all attempts to answer [the question of design] before 1859 are worthless and . . . we will be better off if we ignore them completely'. But Paley, at least, was an Enlightenment thinker, writing a long time after the scientific revolution and 'the last of the magicians' (*sensu* Keynes 1946). His reasoning is not difficult to follow, even if he held some scientific beliefs that are not held today, which is probably the fate of anyone who discusses scientific evidence. Further, as an empiricist and utilitarian, if he were alive today, it would seem very unlikely to find him amongst modern-day proponents of intelligent design. Instead, it is likely that Paley would be delighted over the extent of the understanding of the scientific question of the secondary cause of design in nature, but insistent on reposing the theological question of design about the character of the first cause, and thereby finding the modern grounds to restate his case that the creative forces that designed us express benevolence.

* * *

Perhaps maladaptation presents a more serious problem for Paley's design argument. At seemingly every potential criticism to his argument for the benevolence of God from design, Paley falls back on the weight of observation that the objective of design is never to harm individuals. Maladaptation is *designed* to cause harm; it is its objective. Given the way that Paley used adaptation as evidence for the benevolence of God, a contrasting question provocatively arises: should maladaptation be seen as evidence for the malevolence of God?

Most arguments are won or lost on the premise of question, so there is a need to establish the legitimacy of the contrast. Maladaptation produces a harm that is not necessarily in the same sense as the harm that would jeopardise belief in a benevolent God. Maladaptation causes a decrease in individual fitness, relating to the ability of individual organisms to survive and reproduce. Paley was much more focused on moral value as assessed by promoting pleasure—what, today, might be closer to well-being—rather than survival per se. Indeed, 'that the Deity has superadded pleasure to animal sensations, beyond what was necessary for any other purpose, or when the purpose, so far as it was necessary, might have been effected by the operation of pain' (Paley 1802, pp. 454–455) is the second proposition (alongside beneficial design) upon which Paley rests his argument for divine goodness; albeit that Paley (1802, p. 482) later admits that, in as far as pleasure is functionally necessary for the continued existence of species, it does not prove anything about the character of God beyond beneficial design.

Whilst no one should accept individual fitness as a suitable currency of moral worth for the discussion of well-being (where good would involve increasing fitness and evil decreasing fitness), there is a logical connection between many traits that increase fitness and those that make life easier for individual organisms. Indeed, Paley (1802, pp. 18–25) demonstrates this point for adaptations by analogy between organisms' traits and human tools, such as eyes and telescopes. Moreover, Paley (1802, pp. 511–513) rebukes Hume for overcomplicating that whatever makes for an easier life is an improvement. Consequently, in reverse, traits that decrease fitness might also be associated with traits that inflict greater suffering by making life harder, and in this way detract from well-being. Indeed, this fits with many of the maladaptive traits that have been discussed, from intense sexual rivalry to spermicidal infertility. The point is not to suggest that natural selection furnishing individuals with traits that increase their fitness must always necessarily be good because it depends on the alignment of fitness and well-being; but rather, to suggest that there is a strong case—and perhaps a stronger case than Paley made in reverse—for suggesting that decreasing fitness is usually evil in the sense of entailing a trait that causes some hardship.

In finding other ways to deny the premise, many critics might be tempted to reassert Darwin's (1859, p. 211) claim that 'If a fair balance be struck between the good and evil caused by each part, each will be found on the whole advantageous'. Against the reality of maladaptation, Darwin expresses confidence that any trait would eventually be found to be a beneficial adaptation that contributes to an individual's survival and reproduction. In keeping with such a response, it is recognised that there will always be room to cynically suggest that a trait is an adaptation irrespective of any evidence on its harmfulness because additional data could always cast new light on a current claim (*sensu* Popper 1935). This could apply to any of the traits that have been upheld as examples of maladaptation. But it does not detract from a claim on the basis of current evidence. Further, as well as detailing many traits that would at least have given Darwin pause for thought, in Chapter 3 the logic of natural selection favouring maladaptive traits was expounded. With this in mind, there is little room to be as confident as Darwin was that traits that are favoured by natural selection are necessarily beneficial.

Some critics might be tempted to argue that there is a specific hidden benefit to maladaptation that is being overlooked. In this regard, someone might plausibly suggest that by reducing an individual's ability to survive and reproduce, maladaptation could provide an ecological check to regulate populations of individuals in a beneficial way. In response to the accusation of nature being 'red in tooth and claw', a similar argument was presented for the necessity of the suffering caused by predators and parasites (e.g. Drummond 1894). But, even neglecting to criticise the group selection in this argument, it has misconstrued the nature of maladaptation because these traits could plausibly drive a species to extinction; by contrast, there are no documented cases of a natural predator or parasite driving a species to extinction (that is, excluding anthropogenic examples). Furthermore, as has been discussed in Chapter 4, a potential example of maladaptation that is found to have a 'higher level' benefit for individuals would be better understood as an adaptation.

Other critics might be tempted to make the same argument but with the grander appeal to the link between maladaptation and special features of life that are thought to make living more worthwhile. Such features could range from the supposedly base pleasures like sexual intercourse to the more refined tastes like beauty, which are arguably only appreciated in complex multicellular organisms with eyes, ears, and appropriately schooled brains. The general flaw in this line of reasoning can be most clearly seen in a more patent form of the argument: it could be claimed that human beings would not be the same without maladaptation, where the observed qualities of human beings are held up as having special value. Since the Enlightenment, it has been popular to see the facets of human consciousness in this way, especially amongst free thinkers or humanists (e.g. Harris 2010). To think that it is possible to separate maladaptation from our humanity is certainly to test the imagination, but it is difficult to see how maladaptation improves anything in and of itself. Instead, by making individuals worse at survival and reproduction, maladaptation makes everything else that might also be of value to individuals more of a struggle.

So, if maladaptation is accepted to really be harmful to individual survival and reproduction, it is possible to move beyond the premise of the question to address the question itself. At the outset, it is recognised that there are those that might see a connection between the drivers behind maladaptation and original sin. Arguably, the incentive to compete for a relative advantage over others has the potential to be acted out in the moral theatre to encourage people to do evil to one another. If such an interpretation is pursued in an attempt to make it scientifically respectable, it quickly becomes apparent that the resulting concept of original sin is so very far removed from its theological origins or consequences that the link may as well be ignored altogether.

Once again, Paley (1802) has a pre-emptive response to something like the discovery of maladaptation. Whilst Paley did not find any examples of maladaptive traits, it is clear that he would not have accepted any attribution of malevolence to God. Early on, Paley (1802, p. 77) establishes how each example of adaptation in his design argument provides a separate piece of evidence for the benevolence of God, and so a single example of maladaptation would only provide doubt of benevolent intent for that instance. Overall, then, what really matters for Paley (1802, p. 465) is the relative weight of the instances of good and evil:

> pain, no doubt, and privations exist, in numerous instances, and to a degree, which, collectively, would be very great, if they were compared with any other thing than with the mass of animal fruition. For the application, therefore, of our proposition to that *mixed* state of things which these exceptions induce, two rules are necessary, and both, I think, just and fair rules. One is, that we regard those effects alone which are accompanied with proofs of intention. The other, that when we cannot resolve all appearances into benevolence of design, we make the few give place to many; the little to the great; that we take our judgement from a large and decided preponderancy, if there be one.

So, focusing on those examples where the purpose of design is clear (which, for him, was especially those individual traits with an obvious utility), Paley ultimately suggests that the intention behind design in nature is assessed by weighing the balance of good and evil examples. It is an impossible calculation to actually carry out, and Paley seems to know this, but is not put off because he overwhelmingly finds examples of good design. Paley (1802, p. 468) begs examples where pain or evil is the object of design, finding but two ambiguous cases of venom and predation, which forgoes the need for actual calculation. He discusses each in earnest but, in the end, confidently disregards them on the balance of preponderance that he has found.

With the study of maladaptation, many more examples might be ascribed to the weight of evil in nature, perhaps making it not so clear cut. This is a more general problem with utilitarian philosophy, which extends far beyond the form that Paley espouses. The only way to convincingly persuade that a single answer must be taken is when there is a short-cut that circumvents the need for an actual calculation, just like Paley had done in setting the clear examples of goodness in adaptation against the ambiguous examples of venom and predation.

When considering the weight of adaptation against the weight of mal-adaptation in nature, there is an obvious short-cut that can be taken to determine on what side the total balance rests. To partition any individual's fitness into those traits that enhance or diminish survival and reproduction is, of course, impossible. But, like Paley, perhaps it is possible to forgo an accurate accounting in favour of the fact that individuals have some fitness. If adaptation increases the ability of individuals to survive and reproduce, then it leads each species to abundance; if maladaptation does the reverse, then it leads to extinction. On balance, the continual existence of a species suggests that adaptation outweighs maladaptation, and so it is conceivably the predominant tendency of the creative forces. That the ultimate fate of all species is overwhelmingly likely to be extinction is of little consequence because only those extinctions that are the result of maladaptation are relevant considerations.

Such a calculation begs the question: for the existence of any living thing, isn't it necessarily true that adaptation outweighs maladaptation? No. When considering a particular species, as of yet there are no known safeguards that prevent it from going extinct from maladaptation. That a species is descended from other species is assured, but it is not assured to continue to exist or to be the ancestor of any new species. Further, it could be that a focal species and/or its ancestors owe their existence to chance, despite the predominant tendency being in favour of maladaptation. If this were the case, it might be better to shift focus from the peculiarities that there might be for a single species to the general patterns that emerge for life as a whole. But even at this grandest of scales, it could equally be that there is nothing intrinsic to the evolution of life on Earth that has made it necessary that adaptation outweighs maladaptation, and instead it seems life's persistence owes to good fortune. It is possible, but presumably not very likely. Although it is difficult to be sure, this argument would seem to become increasingly untenable as life continues to exist, which it has on Earth for nearly four billion years. Of course, the persistence of life up until the present may not provide a reliable indication that it will continue to persist into the future. But believing life to be intrinsically heading towards its own destruction would seem to be as foolish as the belief that the sun is not going to rise tomorrow.

If not by chance, there could be a reason for the continuing persistence of life. In Chapter 7, there was a discussion of how a maladaptive barrier must be overcome for life to get started, which may initially bias the balance of adaptation and maladaptation for the earliest life forms towards adaptation. Subsequent evolution may alter this balance in any direction. At present, there is nothing about evolution by natural selection acting through genetics on individuals that is known to bias the balance. Of course, if the balance tipped towards maladaptation, species would be more likely to go extinct—and extinct species cease to give rise to any new species. But to imply a species selection (analogous to natural selection on individuals) would be to suggest that the species that do not go extinct have heritable variation in their balance of adaptive and maladaptive trait opportunities that they can pass on to new species, which remains entirely speculative (as discussed in Chapter 6). Similarly, perhaps there is some higher-order process that shapes the environments that present trait opportunities for species, like the Gaia hypothesis, which leads to an emerging bias in the balance of adaptation and maladaptation. Again, it is interesting speculation, but at present there are no known mechanisms that prevent a species from acquiring maladaptation, thereby risking extinction. When setting the theological question of design against such unknowns, the discovery of such a mechanism would add support to Paley's case, finding a secondary cause that more clearly expresses the benevolence of the first cause; at least, against critics like Gould (2002, p. 127) who argue that natural selection does not tend to assure a benevolent outcome. But, at present, the evidence is lacking, and so all that can be said is that it seems likely that the balance of adaptation and maladaptation predominantly expresses benevolence.

As a final, more general criticism, it has been widely argued, as most famously in Hume's (1739) is–ought problem, that what is cannot infer what should be—or, in other words, what is cannot establish what is good. Whilst there is a general truth here that there is not a necessary connection between existence and goodness, the inference is nonetheless reasonable in the special case when there is a process that determines what is on the basis of its goodness. To the extent that there is an association between fitness and well-being, the cumulative process of natural selection affords some grounds for thinking that existence is not morally random. So, with Darwin (1859, p. 79) during his earlier optimism that still echoed Paley's philosophy, 'we may console ourselves with the full belief, that the war of nature is not incessant, that no fear is felt, that death is generally prompt, and that the vigorous, the healthy, and the happy survive and multiply'.

Therefore, taking the premise that maladaptation challenges the benevolence of design in nature seriously, Paley's reasoning still gives scope for his design argument. Whilst the existence of maladaptation builds the counterargument, there is no reason to think that the weight of evidence has shifted to demolish the argument that design is predominantly benevolent. The evidence is now more mixed than it was at the time of Paley, but the design argument is still tenable. On the basis that life continues to thrive, there is still reason to believe that the instances of good adaptation have outweighed the instances of evil maladaptation, so design still appears overwhelmingly benevolent.

* * *

If not demolished by the discovery of evolution by natural selection or maladaptation, there is a final, more general criticism that is levelled against Paley's design argument. The theologian Alistair McGrath (2008) is highly critical of Paley's design argument, despite being sympathetic to the cause of natural theology in general. McGrath (2008, pp. 97–99) attacks Paley's argument for having no suggestion that design might *emerge* from secondary causes in Paley's description of them, which is arguably true. Indeed, Darwin (1859, pp. 481–482) seems to make a similar point: 'It is so easy to hide our ignorance under such expressions as the "plan of creation", "unity of design", and c., and to think that we give an explanation when we only restate a fact'. McGrath, following Darwin, suggests that Paley's design argument rests on an outlook of an active organism with an essentially static design, which is incoherent after the discovery of the gradual emergence of design through evolution by natural selection. McGrath is therefore criticising Paley's natural theology for its place in an outlook that was radically changed after 1859.

It seems uncharitable to criticise Paley for using language that implicitly assumes a static design as McGrath (2008) does, because he was writing at a time when scientists took the special creation of species for granted. It is a fair criticism of the wider world-view but not of the core argument. Evolution by natural selection could be included among those secondary causes that Paley (1802, p. 420) describes as being created and sustained by God. Yet, there are more substantial problems in Paley's world-view that may tarnish the core argument for modern interpreters. Three interconnected beliefs particularly stand out.

First, although it is not discussed in *Natural Theology*, all of Paley's other works contain references to miracles, especially *Evidences of Christianity* (Paley 1794). For Paley (1794, p. 26), a miracle refers to the way that God 'may interrupt the order which he had appointed', which does not imply a cessation of causality because it follows from 'the volition of the Deity' who is part of the causal structure of reality (if not nature, in its proper sense). Consequently, Paley was not a deist who believed that God made the universe and then left it well alone, which is the only version of theism that receives sympathy from even ardent atheists (*e.g.* Dawkins 2006, pp. 39–40). Miracles are important to Paley because they provide evidence for the divine origin of revelation. Second, Paley (1794, pp. 147–148) went on to summarise: 'If I were to describe in a very few words the scope of Christianity, as a *revelation*, I should say, that it was to influence the conduct of human life by establishing the proof of a future state of reward and punishment, to bring "life and immortality to light"'. Whilst he believed that revelation does pave the way for happiness in this life, Paley (1802, pp. 543–546) thought that revelation established the possibility of eternal happiness in an afterlife, which was obscure from what could be reasoned from nature—hence why 'mankind stood in need of a revelation' (Paley 1794, p. 23). Paley (1794) advocates the truth of Christian revelation on the basis of its historicity and utility.

Third, although it is not explicit, Paley (e.g. 1802, pp. 511–513) seems to follow the Augustinian tradition to argue that free will affords radical indeterminism to the attainment of happiness in the afterlife. In this way, human beings have the ability to act independent of causality, which provides a just basis for reward and punishment in an afterlife. Together, these three points suggest that Paley, like many Enlightenment thinkers and Christian apologists that followed, believed that human activity follows the pattern of divine activity in being set apart from nature, rather than being a part of it.

On this basis, the thrust of McGrath's (2008, pp. 97–99) argument is right that Paley's outlook is of its time, but wrong to suggest that it presents an unresolvable flaw in his reasoning. The issues appear to be largely semantic, revolving around how important theological concepts are conceived (i.e. how older theological concepts are brought into a modern epistemology where 'nature' is a synonym for everything that exists). Such disputes help to separate the expressed world-view from the core argument of Paley's natural theology, which cannot be as easily rejected. Despite clearly having a place in his wider outlook, the core argument in *Natural Theology* is perfectly secular, like the natural theology of the Ancient philosophers. Aspects of Paley's wider outlook rest on the revealed theology of Christianity, but these aspects are peripheral to his design argument. Whilst Paley was not a deist, on its own the core argument in *Natural Theology* is deistic. With miracles holding no place in the core of Paley's design argument, he is really discussing the theological question of design about the character of the first cause—not any secondary causes that may (or may not) stem from divine intervention. Further, with an afterlife only being assured by revelation, Paley does not use natural theology to justify his view of Christianity. In extracting the core of Paley's design argument from the extraneous details of its presentation and his wider outlook, the basic case retains its logical coherence, and remains persuasive to this day.

The neo-Paleyan design argument can be simply put in modern terms. It asks: what is the character of the first cause that has resulted in the world of which we are a part? Based on the traits that evolution by natural selection has tended to furnish us with, it is as if we have been designed with a benevolent intent. Based on imperfections, we cannot say that our design is the best that it possibly could be. But we have every reason to believe, from the security of our provision from nature, that the pursuit of happiness is possible. Again, based on imperfections, we cannot say that our happiness is guaranteed, not least because it ultimately requires us to pursue it. That the first cause is God in the meaning of revealed theology lies beyond the scope of inquiry into natural theology. Instead, natural theology leaves us with the philosophical choice to believe either that the design in the world has arisen by chance, or that there is a benevolent Creator; a choice between coincidence and design. As such, the reasoning of the design argument up until this point simply demonstrates the consistency of nature with a benevolent God as described in revelation—and not an inexorable 'proof' of the existence of this God from what we know of nature. That it is a 'design' argument rather than a 'coincidence' argument rests on the debatable proposition that our provision from nature is sufficient to disregard its origins in chance, but of course both design and coincidence are eminently possible.

For some, this might be a dissatisfying or weak conclusion. Indeed, perhaps it is possible to take natural theology further with secular reasoning (e.g. going deeper into the Augustinian requirements of God as a moral governor in bestowing free will). But for Paley (1802, p. 539), this conclusion is important as a foundation, and perhaps also as a kind of invitation, because

> whereas formerly God was seldom in our thoughts, we can now scarcely look upon any thing without perceiving its relation to him. Every organized natural body, in the provisions which it contains for its sustentation and propagation, testifies a care, on the part of the Creator, expressly directed to these purposes.

At the end of *Natural Theology*, Paley (1802, p. 542) ties this purpose of his natural theology into his wider outlook:

> The existence and character of the Deity, is, in every view, the most interesting of all human speculations. In none, however, is it more so, than as it facilitates the belief of the fundamental articles of *Revelation*. It is a

step to have it proved, that there must be something in the world more than what we see. It is a further step to know, that, amongst the invisible things of nature, there must be an intelligent mind, concerned in its production, order, and support. These points being assured to us by Natural Theology, we may well leave to Revelation the disclosure of many particulars, which our researches cannot reach, respecting either the nature of this Being as the original cause of all things, or his character and designs as a moral governor; and not only so, but the more full confirmation of other particulars, of which, though they do not lie altogether beyond our reasonings and our probabilities, the certainty is by no means equal to the importance.

Paley might not have intended it in this way, but as a foundational philosophy, the design argument provides a basis for a disciplined pluralism. Whilst Paley grounded his own outlook of empirical philosophy and utilitarian ethics on the particulars of his revealed theology (and sometimes used Christian language as a result), his natural theology is deistically untied to a particular religious tradition (insofar as the religion concerns itself with God as the Creator). Consequently, the design argument could be used as a foundation for a diversity of philosophical, theological, and ethical beliefs. In this regard, the strength of the design argument is not that it permits that 'anything goes' in the discovery of those beliefs, but rather that it provides a minimal philosophy, which perhaps protects religious inquiry against the insecurity and dangers of unbridled speculation—or the ethical horrors that can result from the pessimistic degeneration of a philosophy into nihilism.

In the end, then, reconsideration of Paley's natural theology leads to the dismissal of Darwin's (1958, p. 87) self-stated achievement that 'The old argument of design in nature, as given by Paley, which formerly seemed to me so conclusive, fails, now that the law of natural selection has been discovered'. Whilst, for Darwin, the problem of pain came to make him doubt the existence of a benevolent God, this belief should not be accepted to have anything to do with the science of evolutionary biology. The persistence of the atheistic narrative, which describes the foundations of the science of evolutionary biology as coming out of the demolition of Paley's design argument, is a highly dubious account of natural theology and its history. Here, the promotion of an atheist philosophy has overstretched itself to make a demonstrably false claim. As should have always been obvious, good science is secular; beyond the basic philosophy of science about the existence of objective truth, being a scientist does not require us to subscribe to a particular philosophical or ethical doctrine. Indeed, a key strength of science is how its methods can generate objective truth on its own terms, irrespective of the beliefs of those that carried out the science. This enables science to be progressive despite the false theories of previous generations of scientists, and to be done internationally across cultures with different philosophical traditions. There is undoubtedly a historical connection between the rise of science and that of atheism, but atheism is just as dangerously nonsecular as theism is to the free enterprise of science. That said, we should not be unsympathetic to the human need for a philosophy. For all its strengths, science also has its weakness in those meaningful questions about existence that lie beyond the scope of the scientific method. Although it is most satisfactory for science to leave those questions unanswered, we often have a need to find uncertain answers, and so we must turn to the disciplined exercise of faith and reason. Whilst it is difficult to discern a way to a definitive answer to the purpose of design in nature, I hope that we may have a more cognisant debate about the question.

Appendix

A simple mathematical model
of maladaptation

Chapter 3 has provided a verbal argument for the possibility of maladaptation. Some expert readers may prefer to see the verbal logic made crystal clear in a mathematical model. The purpose of such a model is not to demonstrate the coherence of a maladaptive hypothesis for any specific example, such as those in Chapters 4 and 5, but rather to provide a general outline of the key properties of maladaptation that have been identified. In particular, a basic population genetic and ecological approach can show that: (1) maladaptation is not readily favoured by natural selection under standard assumptions, (2) maladaptation is made possible by incorporating an additional ecological effect, (3) maladaptation can generate unequal payoffs in the genome that can escape suppression (or resistance), and (4) maladaptation can lead to population extinction.

As there is already enough complexity to consider, classic simplifying assumptions of population genetic models are made. Briefly, these include haploidy, panmixia, and nonoverlapping generations, which are widely regarded as permissible for the construction of a general model because they rest on clear biological premises with known consequences (Crow and Kimura 1970, pp. 173–174). Deviating from the classic approach, the model of maladaptation is necessarily eco-evolutionary (*sensu* Hendry 2016) in considering both absolute and relative fitness, wherein population genetics can be used to describe the effect of selection, and population ecology can be used to describe its effect on population size.

To begin, consider an allele for a maladaptive trait (M) that is competing against an alternative wild-type allele (m) at a locus. Using the framework of inclusive fitness where maladaptation is like spite (Hamilton 1970), the relative fitness effects of the allele for maladaptation can be described by a cost to the actor ($-c$) and a benefit to the recipient (which is described as $-b$ because of a focus on the case of harm when this benefit is negative). To calculate the relative fitness of the allele for maladaptation, the cost and negative benefit must be weighted by a probability (p_c and p_b respectively) that those relative fitness effects return to the allele for maladaptation (somewhat like an F-statistic that is commonly used to describe spatially explicit population structures; cf. Wright 1951), which relate to the allele that is carried by the actor and recipient within the population structure of the interaction. Under the assumption of weak selection (in keeping with classical inclusive fitness theory; Grafen 1984), relative fitness of the allele for maladaptation can be calculated by the addition of these fitness effects:

$$\omega_M = 1 - p_b b - p_c c. \tag{1}$$

For most traits under consideration, the actor will always receive the cost of maladaptation (i.e. $p_c = 1$) because it is exhibits a trait that causes harm to the recipient; however it is possible for an actor to make the group pay the cost of maladaptation (e.g. through paying with a public good), and so for generality the extra probability term is left in the model.

The relative fitness of the wild-type allele depends on the frequency at which it interacts with the allele for maladaptation. The interaction frequency is the product of the frequency of the allele for maladaptation (f) and the complementary probability that the allele for maladaptation interacts with the wild-type allele ($1 - p_b$, $1 - p_c$). The interaction with the allele for maladaptation is not the only interaction that the wild-type allele could have, as it could also interact with itself (whereupon it does not display the maladaptive trait; i.e. has a relative fitness of 1). The mean relative fitness of the wild-type allele across interaction partners can be calculated by dividing the interaction frequency by the frequency of the wild-type allele ($1 - f$):

$$\omega_m = 1 - \left(\frac{f(1-p_b)}{1-f}\right) b - \left(\frac{f(1-p_c)}{1-f}\right) c. \tag{2}$$

To be favoured by natural selection, the allele for maladaptation must have greater relative fitness than the competing wild-type allele ($\omega_M > \omega_m$), which simplifies to:

$$-\left(\frac{p_b - f}{p_c - f}\right) b - c > 0. \tag{3}$$

For there to be any possibility of natural selection favouring the allele for maladaptation, one of the probabilities must be lower than the frequency of the allele for maladaptation.

The key scenario to consider is what would happen at invasion, when the allele for maladaptation first arises by mutation. When doing so, it is worth bearing in mind that the probabilities may take some time to establish themselves within a population structure, not least because a single mutation may not be shared by any individual if it has only just spontaneously occurred; for population structures where this matters (which is not all conceivable population structures), the assumption of weak selection permits the allele for maladaptation to spread to an equilibrium within the population structure by random drift before evaluating the direction of selection. Regardless, considering invasion using the inequality as given, the frequency of the allele for maladaptation would be very small, which makes it very unlikely that the frequency would exceed the probabilities. Further, if the often biologically sound assumption that the cost is only paid by the allele for maladaptation is made, even if the harm of the negative benefit is only returned to the wild-type allele (i.e. $p_b = 0$), the cost to itself is overwhelmingly likely to be outweighed by the harm to others; this makes good sense because the cost is paid by every allele for maladaptation, but the harm only occurs to a small fraction of the wild-type alleles that interact with the allele for maladaptation, which occurs very infrequently when the allele for maladaptation is rare. Therefore, without additional factors in the model, it cannot be suggested that maladaptation would readily be favoured by natural selection. Clarifyingly, this makes clear that the evolution of maladaptation is not an 'obvious' outcome—at least, not from the standard assumptions of population genetic models (cf. Wade 1980).

In Chapter 3, the proposed solutions are two-fold. First, there is an evolutionary effect in small populations, where smaller populations may afford a mutant allele a higher frequency by the simple fact of being counted among fewer individuals (Hamilton 1971). On its own, this idea is of little importance because populations would have to be very small for substantial frequency inflation to occur, whereupon populations are liable to go extinct anyway. But alongside the idea of localised competition, the evolutionary effect may be more influential because small arenas of competition may arise in large, viable populations (Gardner and West 2004). Second, there is an ecological effect from the scale of competition due to density-dependent feedback or similar (Taylor 1992). In the preceding model, it was implicitly assumed that competition occurs at the level of the whole population. When the actor harms the recipient, both the actor and the recipient have reduced fitness, but other individuals elsewhere in the population obtain a relative fitness benefit (Lehmann et al. 2006). Indeed, this is why frequency is so important to the direction of selection: because the alleles benefit through individuals elsewhere in the population in accordance with their allele frequency. Alternatively, if competition occurs at the level of the group, then other members of the group may receive that relative fitness benefit.

The ecological effect can be brought into the model using an additional term for density-dependent feedback (d). Like the others, this fitness effect is returned to the allele for maladaptation in accordance with the probability of the allele within the group (p_d). The new relative fitness equations for the alleles are:

$$\omega_M = 1 - p_b b - p_c c + p_d d, \tag{4}$$

$$\omega_m = 1 - \left(\frac{f(1 - p_b)}{1 - f}\right) b - \left(\frac{f(1 - p_c)}{1 - f}\right) c + \left(\frac{f(1 - p_d)}{1 - f}\right) d. \tag{5}$$

Technically speaking, to maintain the precise definitions of the terms of inclusive fitness, the density-dependent feedback must arise for individuals other than the actor and recipient (Patel et al. 2020), which is not strictly necessary here. The condition for the allele for maladaptation to be favoured by natural selection is:

$$-\left(\frac{p_b - f}{1 - f}\right) b - \left(\frac{p_c - f}{1 - f}\right) c + \left(\frac{p_d - f}{1 - f}\right) d > 0. \tag{6}$$

This equation would seemingly present the same difficulties for an invading mutant as before, but the density-dependent feedback term (d) needs to be interpreted. One way to conceptualise density-dependent feedback is through a familiar expression (May 1976), where the fitness effect is dependent upon the proximity of the focal population size (n) to the focal carrying capacity (k). It is also necessary to consider any reduction in group population owing to maladaptation, which would depend on mean fitness ($\bar{\omega}$) prior to this fitness effect being accounted:

$$d = 1 - \frac{n\bar{\omega}}{k}. \tag{7}$$

In principle, this equation is ambivalent to the scale of competition. If it occurred at the level of the whole population, it would not impact relative fitness because it would be the same for all individuals. If it occurs at a smaller scale like at the level of the group, then it does impact relative fitness because the density-dependent feedback non-randomly benefits group members. Most straightforwardly (Queller 1994), the population would already be structured into patches where the group population size is already at the group carrying capacity, and so the equation simplifies to the complement of group mean fitness ($1 - \bar{\omega}$). For the groups with an allele for maladaptation, the group mean fitness is:

$$d = b + c. \tag{8}$$

Substituting this into the condition for the allele for maladaptation to be favoured by natural selection, a familiar form of Hamilton's rule can be derived for natural selection to favour the allele for maladaptation:

$$-\left(\frac{p_b - p_d}{p_c - p_d}\right) b - c > 0. \tag{9}$$

The multiplier of the benefit is known as relatedness (r). If the key scenario of invasion is once again considered, the frequency terms vanish for this new inequality, and everything depends on the interpretation of the probabilities.

Interpreting the probabilities is not always straightforward. One general approach is to consider assortment factors, which describe a fixed probability of allelic interaction (Hamilton 1971; see also Madgwick 2020). Assortment factors can be positive or negative, but are always bound between zero and one. A positive assortment factor describes the probability that an allele interacts with a copy of itself, which most importantly arises in kin-structured populations by common ancestry. For example, sibling interactions take place with them sharing 50% of their alleles through descent by having parents in common, and so this interaction would be described by a positive assortment factor of $a^+ = 0.5$. If siblings do not share alleles that are copies of the same allele in a parent, then they would share alleles by random chance in accordance with the frequency of the allele in the population, assuming that parents are outbred ($(1 - a^+)f$). As such the probability of an allele within each role can be replaced by a positive assortment factor as the sum of these two sources of allele sharing:

$$p = a^+ + \left(1 - a^+\right)f. \tag{10}$$

By contrast, a negative assortment factors describes the probability that an allele interacts with a copy that is not itself, which most importantly arises in populations that are structured by genetic recognition. For example, an in-group member identifies an out-group member that does not display a trait (e.g. an odour), but it only has a 75% success rate of correctly identifying out-group members (i.e. a 25% false-positive rate), and so this interaction would be described by a negative assortment factor of $a^- = 0.75$. Like before, if in-group members do not correctly identify an out-group member, individuals share alleles by random chance in accordance with the frequency of the allele in the population $((1 - a^-)f)$. So, the probability of an allele within each role can be replaced by a negative assortment factor from this second source of sharing alleles only:

$$p = \left(1 - a^-\right)f. \tag{11}$$

It is worthy of note that, if negative assortment occurred, the biological relevance of the assumptions of the inclusive fitness may break down because it implicitly assumes 'one-way assortment' (Madgwick 2020), whereby an actor always carries out its action but an individual may be a recipient many times over. This is most relevant for when the allele for maladaptation obtains higher frequencies, and the wild-type allele becomes rarer (and therefore the recipient of multiple interactions). For many interactions, it is more relevant to consider 'two-way assortment', whereby an actor and recipient may only play their role once, which is more like a situation where individuals are grouped together (e.g. paired up) in a population structure. It is possible to apply two-way assortment to the framework of inclusive fitness, but it becomes substantially more complicated. As negative assortment is not especially relevant here (for reasons that will become apparent), this issue is neglected for simplicity.

If the classic assumption that an actor is the sole payer of the cost (i.e. $p_c = 1$) is made, the four possible combinations of positive and negative assortment factors for the probability of the allele for maladaptation interacting with itself in a recipient and the group can be substituted into Hamilton's rule (Table A1). Given that the assortment factors are constrained to have a positive value between zero and one, and that the allele for maladaptation requires relatedness to be negative to be favoured by natural selection, the possibility of maladaptation under different assortment factor combinations can be assessed. The results reveal that negative relatedness is only appreciable at invasion when the probability of the allele for maladaptation interacting with itself in the group has positive assortment. Negative relatedness is often frequency dependent, but at invasion the frequency of the allele for maladaptation is close to zero, and so the invasion criterion can be interpreted as if it is frequency independent; this removes the possibility of negative assortment of alleles in the group contributing to negative relatedness because negative assortment is inherently frequency dependent. Positive assortment of alleles in the group gives the scope for appreciable negative relatedness at invasion irrespective of the (positive or negative) assortment of the allele for maladaptation in recipients. If both probabilities of the allele for maladaptation interacting with itself in the recipient and the group have positive assortment, negative relatedness can be completely frequency independent, which is a scenario of special interest in describing the interaction among different classes of kin (e.g. siblings and cousins). Therefore, the substitution of assortment factors shows that it is possible for natural selection to favour maladaptation, which is most likely when there is the positive assortment of the beneficiaries of the ecological effect in the group (or locality).

Table A1 Impact of positive (+) and negative (−) assortment of the probability of sharing alleles within a role (p_b, p_d) on relatedness (r), and the scope that this affords an appreciable level of negative relatedness at invasion for natural selection to favour an allele for maladaptation.

p_b	p_d	r	Appreciable negative relatedness?
+	+	$\dfrac{a_b^+ - a_d^+}{1 - a_d^+}$	Yes, when $a_d^+ > a_b^+$.
+	−	$\dfrac{a_b^+ - \left(a_b^+ - a_d^-\right)f}{1 - \left(1 - a_d^-\right)f} \approx a_b^+$	No, impossible.
−	+	$\dfrac{-a_d^+ - \left(a_b^- - a_d^+\right)f}{\left(1 - a_d^+\right)(1 - f)} \approx \dfrac{-a_d^+}{1 - a_d^+}$	Yes, when $a_d^+ > 0$.
−	−	$\dfrac{-\left(a_b^- - a_d^-\right)f}{1 - \left(1 - a_d^-\right)f} \approx 0$	No, as too small when $f \approx 0$.

Whilst this analysis of relatedness is informative, when considering maladaptation, there must be a reduction in absolute fitness. At this point, the analysis of maladaptation deviates from the analysis of spite. It is noteworthy that spite may be an adaptation if it increases absolute fitness even though it harms both the actor and recipient, which is possible when there is a local increase in density ($d > b + c$). For maladaptation, there must be a local decrease in density, which may arise from density-dependent feedback that does not fully compensate the fitness loss from maladaptation at the level of the group ($d < b + c$). This takes us back to equation (6), before the substitution of fully compensatory density-dependent feedback, which leads to simplified results in terms of relatedness (r). Without this substitution, the broad results of the frequency independence of negative relatedness remain the same, even if they are framed in a less familiar way without a single relatedness term. When a probability of interacting with an allele for maladaptation is interpreted with a positive assortment factor, it simplifies to equal the assortment factor:

$$\left(\frac{p - f}{1 - f}\right) = \left(\frac{a^+ + (1 - f)\,a^+ - f}{1 - f}\right) = a^+. \tag{12}$$

If all alleles in the different roles are positively assorted (i.e. among different classes of kin), this still gives scope for the condition for natural selection to favour the allele for maladaptation to be frequency independent. But, more generally, positive assortment of alleles in the group (a_d^+) provides the same scope for natural selection to favour the mutant allele for maladaptation at invasion irrespective of the (positive or negative) assortment of recipients. So the key result of the analysis holds even when the calculation of a single relatedness term becomes difficult, as it does for maladaptation. Therefore, in contrast to the previous model, it is evident that maladaptation is made possible by incorporating the ecological effect (d) because it permits the allele for maladaptation to be favoured at invasion when it is initially a rare mutation.

Without considering a multilocus model, the calculations of relatedness demonstrate the capacity for maladaptation to generate unequal payoffs for genes throughout the genome, but also why such asymmetries may persist. An allele for maladaptation and an unlinked allele elsewhere in the genome may have different relative fitness effects from maladaptation because of different allele frequencies and assortment factors. Consider a mutation at an unlinked locus that forms a new suppressor allele. By its rarity, it is unlikely to have appreciable negative relatedness unless there is negative assortment of the group. The assortment factor for the group is likely to be shared for the allele for maladaptation and the suppressor allele when it arises from population structure. The assortment factor for the recipients may not be shared with the two alleles, especially when the maladaptation relies upon genetic recognition to identify recipients that are less likely to carry the allele for maladaptation. In this case, the assortment factor for the recipients may well be closer to zero (or the assortment factor for the group). There may still be negative relatedness for the suppressor allele when recipients are negatively assorted because of its low frequency. Consequently, a suppressor may be unable to spread through a population and/or have its spread curtailed at higher frequencies, which may enable asymmetric payoffs to persist across the genome.

The final property to demonstrate is that it is conceivable that maladaptation could lead to population extinction. This is much more difficult to show because it depends on the incorporation of relative fitness into an eco-evolutionary model of absolute fitness that relates to population size. A simple model can be used to demonstrate this basic property.

To start with, it is worth considering the construction of the relative fitness equations in a way that is ambivalent to weak selection because one plausible way for maladaptation to drive extinction is as a trait of major effect. The relative fitness equations can be made ambivalent through the multiplication of fitness effects, which is the standard approach in population genetics to make them statistically independent (cf. Wright 1945). The core insights of inclusive fitness theory are robust to strong selection, but there may be complications when the social behaviour is expressed prior to the diffusion of the allele through the population structure (i.e. when a new mutation is not found in other individuals) because such an allele may be more likely to suffer strong selection against it before it becomes established in the population structure; the subsequent analysis using the framework of inclusive fitness assumes that the probabilities of interaction are defined by constant assortment factors of the population structure, which may not make for a biologically relevant model when this is not the case. To be ambivalent, the multiplication of fitness effects yields:

$$\omega_M = (1 - p_b b)(1 - p_c c)(1 + p_d d), \tag{13}$$

$$\omega_m = \left(1 - \left(\frac{f(1 - p_b)}{1 - f}\right)b\right)\left(1 - \left(\frac{f(1 - p_c)}{1 - f}\right)c\right)\left(1 + \left(\frac{f(1 - p_d)}{1 - f}\right)d\right). \tag{14}$$

There is no simple derivation of Hamilton's rule for the condition that natural selection favours the allele for maladaptation.

Relative fitness refers to the fitness effects that arise from the locus for maladaptation, depending on which allele an individual has. It is noteworthy that the ecological effect (d) brings a quantitative degree of 'soft selection' (*sensu* Wallace 1975) into the model through the density-dependent feedback compensating members of the group for the fitness loss from maladaptation. Paradigmatic soft selection would occur when the mean relative fitness of the group is unchanged by the evolution of maladaptation but, of course, this does not describe maladaptation (which is defined by some level of fitness loss). Paradigmatic 'hard selection' would occur as in the first model that was discussed where there is no ecological effect ($d = 0$) to compensate the fitness loss (and so it reduces the probability of survival). Hard selection is not conducive to the evolution of maladaptation, which instead requires some level of soft selection to give a frequency-independent criterion for the invasion of the allele for maladaptation.

By contrast, absolute fitness is the product of all the relative fitness effects from all loci, which captures the genetic load on the population from hard selection. As those loci are not of special interest, the total fitness loss from independent loci elsewhere in the genome can be captured by a combined fitness effect $(1 - e)$. As such, the mean absolute fitness of the whole population can be calculated through the sum of fitness effects at each locus, for example,. for the locus of maladaptation: $\bar{\omega} = f\omega_M + (1 - f)\omega_m)$ and for the product of each independent locus:

$$\bar{w} = (1 - fb)(1 - fc)(1 + fd)(1 - e). \tag{15}$$

The mean absolute fitness is important to the population dynamics of all the groups in the population. The ecological effect (d) can be conceptualised as a reduction in density-dependent mortality that compensates the members of the group for the fitness loss from maladaptation *on their particular patch* (i.e. how full each patch is); indeed, a constant assortment factor in probability of the allele within the group (p_d) assumes that the population structure is the same irrespective of any changes in the population density. As a result, there is a need to incorporate an entirely separate ecological effect of density-dependent feedback that occurs at the level of the whole population, which determines how maladaptation impacts the size of the whole population across all the viable patches of the population's habitat. This ecological effect can be conceptualised as how many of those patches are filled.

The ecological effect of density-dependent feedback on the whole population can describe how the population size (N) changes between the generations using a discrete-time logistic equation (May 1976). Absolute fitness can be incorporated into this equation (cf. Madgwick et al. in press) to describe how maladaptation would impact population dynamics in terms of the Malthusian parameter describing the intrinsic rate of population increase (m; often referred to as r in ecological models) and the population carrying capacity (K) describing the size of the ecological niche:

$$N' = N\bar{w}m\left(1 - \frac{N\bar{w}}{K}\right). \tag{16}$$

Even though this equation is simple, it can exhibit complex dynamics, such as chaos (when $3.5 < \bar{w}m < 4$ where 4 is the maximum value of stable dynamics).

The carrying capacity is not the equilibrium population size in this model. In the absence of maladaptation (i.e. $f = 0$), the equilibrium size of the population (K_e) depends on the absolute fitness of individuals ($\bar{w} = 1 - e$):

$$K_e = K \left(\frac{m\bar{w} - 1}{m\bar{w}^2} \right). \tag{17}$$

With the spread of the allele for maladaptation, the equilibrium population size changes. Interestingly, despite the simplicity of the discrete-time logistic equation, this can lead to alternative outcomes for the new equilibrium population size. The typical case would be that maladaptation, which reduces absolute fitness, decreases the equilibrium population size. In this case, there is no reason to think that it inexorably leads the population to extinction (*contra* Bell 2017), as it may just lead to a lower but also stable equilibrium. Alternatively, it is possible for maladaptation, which still reduces absolute fitness, to increase the equilibrium population size when mean absolute fitness is greater than a critical value:

$$\bar{w}^* > \frac{m\bar{w} - \sqrt{m^2\bar{w}^2 - 4\,(m\bar{w} - 1)}}{2\,(m\bar{w} - 1)}. \tag{18}$$

This inequality is satisfied for larger Malthusian parameters ($m > 2$) and smaller reductions in a high level of absolute fitness (i.e. $-b - c + d$ and e are closer to zero). The disassociation between absolute fitness and population is an important clarification of the interpretation of maladaptation; whilst it is reasonable to suggest that a decrease in absolute fitness may often be associated with a decrease in population, the relationship is not trivial. As a result, it is problematic to infer maladaptation on the basis of a decrease in the size of the whole population, which should instead focus on the decrease in the group population. Fortunately, this evidence is easier to obtain.

In this eco-evolutionary model, evolving populations can go extinct. It is possible for the increasing frequency of maladaptation to tip the balance of the rates of population change so that the equilibrium population size is less than zero ($K_e < 0$). To be viable before maladaptation ($K_e > 0$), the mean absolute fitness in the absence of maladaptation ($\bar{w} = 1 - e$) is constrained:

$$(1 - e)\,m > 1. \tag{19}$$

This simply expresses that the maximum rate of net reproduction after survival exceeds the level required for population replacement. By breaking the mean absolute fitness of the population into the relative fitness of the locus for maladaptation and the other independent loci in the genome (as in equation 15), the condition for population extinction can be expressed in terms of the mean relative fitness at the locus for maladaptation:

$$(1 - fb)\,(1 - fc)\,(1 + fd) < \frac{1}{m\,(1 - e)}. \tag{20}$$

Therefore, it is conceivable that natural selection, by favouring the allele for maladaptation, drives the population extinct by increasing the relative fitness of an allele at its locus at the expense of the absolute fitness of the individual, and the survival of the whole population. As the model already includes population structure through the use of constant assortment factors in the probabilities of interaction, this can occur in as far as the change in population size does not influence the population structure; in other words, this would occur in as far as density-dependent feedback on the whole population changes the number of patches that are filled by a group rather than how full each group is on a patch. Maladaptation is clearly a very different method of extinction than has been conceived by adaptation (or non-adaptation) because it does not involve a discontinuous transition in the population equilibrium across Malthusian parameters (Gyllenberg et al. 2002; see also Webb 2003).

There is much more that could be done to develop the model of population genetics and ecology to understand the evolution of maladaptation, but the sketch that is presented here provides a general outline of the key properties of maladaptation that have been identified. The model shows that: (1) maladaptation is not readily favoured by natural selection under standard assumptions, (2) maladaptation is made possible by incorporating an additional ecological effect, (3) maladaptation can generate unequal payoffs in the genome that can escape suppression (or resistance), and (4) maladaptation can lead to population extinction.

Bibliography

Abbott J. K., A. K. Nordén, and B. Hansson, (2017). 'Sex chromosome evolution: historical insights and future perspectives'. *Proc. R. Soc. B Biol. Sci.* 284. https://doi.org/10.1098/rspb.2016.2806

Abegglen L. M., A. F. Caulin, A. Chan, et al., (2015). 'Potential mechanisms for cancer resistance in elephants and comparative cellular response to DNA damage in humans'. *JAMA* 314: 1850–1860. https://doi.org/10.1001/jama.2015.13134

Ågren J. A., and A. G. Clark, (2018). 'Selfish genetic elements'. *PLoS Genet.* 14: e1007700. https://doi.org/10.1371/journal.pgen.1007700

Ågren J. A., M. Munasinghe, and A. G. Clark, (2019). 'Sexual conflict through mother's curse and father's curse'. *Theor. Popul. Biol.* 129: 9–17. https://doi.org/10.1016/j.tpb.2018.12.007

Aktipis C. A., A. M. Boddy, G. Jansen, et al., (2015). 'Cancer across the tree of life: cooperation and cheating in multicellularity'. *Philos. Trans. R. Soc. B Biol. Sci.* 370: 20140219–20140219. https://doi.org/10.1098/rstb.2014.0219

Alabi T., J. P. Michaud, L. Arnaud, et al., (2008). 'A comparative study of cannibalism and predation in seven species of flour beetle'. *Ecol. Entomol.* 33: 716–726. https://doi.org/10.1111/j.1365-2311.2008.01020.x

Alexander R., and G. Borgia, (1978). 'Group selection, altruism, and the levels of organization of life'. *Annu. Rev. Ecol. Syst.* 9: 449–474. https://doi.org/10.1146/annurev.es.09.110178.002313

Alexander R. D., (1979 [1982]). *Darwinism and Human Affairs*. University of Washington Press, London.

Allison A. C., (1954). 'Protection afforded by sickle-cell trait against subtertian malarial infection'. *Br. Med. J.* 1: 290–294. https://doi.org/10.1136/bmj.1.4857.290

Alvarado-Cárdenas L. O., E. Martínez-Meyer, T. P. Feria, et al. (2013). 'To converge or not to converge in environmental space: testing for similar environments between analogous succulent plants of North America and Africa'. *Ann. Bot.* 111: 1125–1138. https://doi.org/10.1093/aob/mct078

Amundson R., (1996). 'Historical development of the concept of adaptation', pp. 11–54 in *Adaptation*, edited by Rose M. R., Lauder G. V., Academic Press, San Diego.

Anderson S., A. T. Bankier, B. G. Barrell, et al., (1981). 'Sequence and organization of the human mitochondrial genome'. *Nature* 290: 457–465. https://doi.org/10.1038/290457a0

Anderson R. M., and R. M. May, (1991 [2010]). *Infectious Diseases of Humans: Dynamics and Control.* Oxford University Press, New York.

Andersson M., (1982). 'Female choice selects for extreme tail length in a widowbird'. *Nature* 299: 818–820. https://doi.org/10.1038/299818a0

Aquinas T., (1259 [1998]). *Thomas Aquinas: Selected Writings* (R. McInerny, Ed.). Penguin Books, London.

Ardlie K. G., and L. M. Silver, (1996). 'Low frequency of mouse t haplotypes in wild populations is not explained by modifiers of meiotic drive'. *Genetics* 144: 1787–1797. https://doi.org/10.1093/genetics/144.4.1787

Ardlie K. G., and L. M. Silver, (1998). 'Low frequency of t haplotypes in natural populations of house mice (Mus musculus domesticus)'. *Evolution* (N Y) 52: 1185–1196. https://doi.org/10.1111/j.1558-5646.1998.tb01844.x

Augustine, 467 ([2003]). *City of God* (H. Bettenson, and G. R. Evans, eds.). Pelican Books, London [Penguin Books, London].

Aumer D., E. Stolle, M. Allsopp, et al., (2019). 'A single SNP turns a social honey bee (Apis mellifera) worker into a selfish parasite'. *Mol. Biol. Evol.* 36: 516–526. https://doi.org/10.1093/molbev/msy232

Bartz S. H., and B. Hölldobler, (1982). 'Colony founding in *Myrmecocystus mimicus* Wheeler (Hymenoptera, Formicidae) and the evolution of foundress associations'. *Behav. Ecol. Sociobiol.* 10: 137–147. https://doi.org/10.1007/BF00300174

Bauer H., J. Willert, B. Koschorz, et al., (2005). 'The t complex-encoded GTPase-activating protein Tagap1 acts as a transmission ratio distorter in mice'. *Nat. Genet.* 37: 969–973. https://doi.org/10.1038/ng1617

Beebe N. W., D. Pagendam, B. J. Trewin, et al., (2021). 'Releasing incompatible males drives strong suppression across populations of wild and Wolbachia-carrying *Aedes aegypti* in Australia'. *Proc. Natl. Acad. Sci. U. S. A.* 118. https://doi.org/10.1073/pnas.2106828118

Beeman R. W., K. S. Friesen, and R. E. Denell, (1992). 'Maternal-effect selfish genes in flour beetles'. *Science* 256: 89–92. https://doi.org/10.1126/science.1566060

Belcher L. J., P. G. Madgwick, S. Kuwana, et al., (2022). 'Developmental constraints enforce altruism and avert the tragedy of the commons in a social microbe'. *Proc. Natl. Acad. Sci. U. S. A.* 119: 1–10. https://doi.org/10.1073/pnas.2111233119

Bell G., (2017). 'Evolutionary rescue'. *Annu. Rev. Ecol. Evol. Syst.* 48: 605–627. https://doi.org/10.1146/annurev-ecolsys-110316-023011

Belov K., (2012). 'Contagious cancer: lessons from the devil and the dog'. *Bioessays* 34: 285–292. https://doi.org/10.1002/bies.201100161

Benabentos R., S. Hirose, R. Sucgang, et al., (2009) 'Polymorphic members of the lag gene family mediate kin discrimination in Dictyostelium'. *Curr. Biol.* 19: 567–572. https://doi.org/10.1016/j.cub.2009.02.037

Bernasconi G., and J. E. Strassmann, (1999). 'Cooperation among unrelated individuals: the ant foundress case'. *Trends Ecol. Evol.* 14: 477–482. https://doi.org/10.1016/S0169-5347(99)01722-X

Biebricher C. K., M. Eigen, and R. Luce, (1981). 'Product analysis of RNA generated de novo by Qβ replicase'. *J. Mol. Biol.* 148: 369–390. https://doi.org/10.1016/0022-2836(81)90182-0

Biernaskie J. M., S. A. West, and A. Gardner, (2011) 'Are greenbeards intragenomic outlaws?' *Evolution* (NY) 65: 2729–2742. https://doi.org/10.1111/j.1558-5646.2011.01355.x

Biernaskie J. M., A. Gardner, and S. A. West, (2013) 'Multicoloured greenbeards, bacteriocin diversity and the rock-paper-scissors game'. *J. Evol. Biol.* 26: 2081–2094. https://doi.org/10.1111/jeb.12222

Birch J., (2012). *Social Revolution: Review essay on Andrew F. G. Bourke: Principles of Social Evolution*, Oxford University Press, Oxford, 2011. Biol. Philos. 27: 571–581. https://doi.org/10.1007/s10539-011-9300-4

Boddy A. M., H. Kokko, F. Breden, et al., (2015). 'Cancer susceptibility and reproductive trade-offs: a model of the evolution of cancer defences'. *Philos. Trans. R. Soc. B Biol. Sci.* 370. https://doi.org/10.1098/rstb.2014.0220

Bonner J. T., (1974). *On Development: The Biology of Form.* Harvard University Press, Cambridge, MA.

Bonner J. T., (1988). *The Evolution of Complexity by Means of Natural Selection.* Princeton University Press, Princeton, NJ.

Borrett A., (2021). 'How UK house prices have soared ahead of average wages'. *New Statesman.* https://www.newstatesman.com/politics/2021/05/how-uk-house-prices-have-soared-ahead-average-wages

Botha M. C., and P. Beighton, (1983). 'Inherited disorders in the Afrikaner population of southern Africa. Part I. Historical and demographic background, cardiovascular, neurological, metabolic and intestinal conditions'. *South African Med. J.* 64: 609–612.

Bourke A. F. G., (2011). *Principles of Social Evolution.* Oxford University Press, New York.

Bowen N. J., I. K. Jordan, J. A. Epstein, et al., (2003). 'Retrotransposons and their recognition of pol II promoters: a comprehensive survey of the transposable elements from the complete genome sequence of Schizosaccharomyces pombe'. *Genome Res.* 13: 1984–1997. https://doi.org/10.1101/gr.1191603

Bowler P. J., (1983). *The Eclipse of Darwinism: Anti-Darwinian Evolution Theories in the Decades Around 1900.* The Johns Hopkins University Press, Baltimore.

Boyd R., and P. J. Richerson, (1985). *Culture and the Evolutionary Process.* University of Chicago Press, Chicago.

Bradshaw A. D., (1991). 'The Croonian Lecture, 1991: genostasis and the limits to evolution'. *Philos. Trans. R. Soc. B* 333: 289–305. https://doi.org/10.1098/rstb.1991.0079

Brady S. P., D. I. Bolnick, A. L. Angert, et al., (2019a). 'Causes of maladaptation'. *Evol. Appl.* 12: 1229–1242. https://doi.org/10.1111/eva.12844

Brady S. P., D. I. Bolnick, R. D. H. Barrett, et al., (2019b). 'Understanding maladaptation by uniting ecological and evolutionary perspectives'. *Am. Nat.* 194: 495–515. https://doi.org/10.1086/705020

Brelsfoard C. L., and S. L. Dobson, (2009). 'Wolbachia-based strategies to control insect pests and disease vectors'. *Asia-Pacific J. Mol. Biol. Biotechnol.* 17: 55–63. https://doi.org/10.1007/978-0-387-78225-6_9

Browne J., (1995). *Charles Darwin: Voyaging* (volume 1). Random House, London.

Bull J. J., (2017). 'Lethal gene drive selects inbreeding'. *Evol. Med. Public Heal.* 2017: 1–16. https://doi.org/10.1093/emph/eow030

Bull J. J., and H. S. Malik, (2017). 'The gene drive bubble: new realities'. *PLoS Genet.* 13: e1006850. https://doi.org/10.1371/journal.pgen.1006850

Bull J. J., C. H. Remien, and S. M. Krone, (2019). 'Gene-drive-mediated extinction is thwarted by population structure and evolution of sib mating'. *Evol. Med. Public Heal.* (2019): 66–81. https://doi.org/10.1093/emph/eoz014

Burt A., (2003). 'Site-specific selfish genes as tools for the control and genetic engineering of natural populations'. *Proc. R. Soc. B Biol. Sci.* 270: 921–928. https://doi.org/10.1098/rspb.2002.2319

Burt A., and R. Trivers, (2006 [2008]). *Genes in Conflict*. Harvard University Press, Cambridge, MA.

Buss L. W., (1987). *The Evolution of Individuality*. Princeton University Press, Princeton, NJ.

Buttery N. J., C. N. Jack, B. Adu-Oppong, et al., (2012) 'Structured growth and genetic drift raise relatedness in the social amoeba *Dictyostelium discoideum*'. *Biol. Lett.* 8: 794–797. https://doi.org/10.1098/rsbl.2012.0421

Cairns-Smith A. G., (1982). *Genetic Takeover: And the Mineral Origins of Life*. Cambridge University Press, Cambridge.

Cairns-Smith A. G., (1985). *Seven Clues to the Origin of Life*. Cambridge University Press, Cambridge.

Calvin M., (1969). Chemical Evolution: molecular evolution towards the origin of living systems on earth and elsewhere. Oxford University Press, New York.

Carneiro M., D. Hu, J. Archer, C. et al., (2017). 'Dwarfism and altered craniofacial development in rabbits is caused by a 12.1 kb deletion at the HMGA2 locus'. *Genetics* 205: 955–965. https://doi.org/10.1534/genetics.116.196667

Carroll L. S., S. Meagher, L. Morrison, et al., (2004). 'Fitness effects of a selfish gene (the Mus t complex) are revealed in an ecological context'. *Evolution* (NY) 58: 1318–1328. https://doi.org/10.1111/j.0014-3820.2004.tb01710.x

Carvalho M. De, G. S. Jia, A. N. Srinivasa, et al., (2022). 'The wtf meiotic driver gene family has unexpectedly persisted for over 100 million years'. *Elife* 11: 1–41. https://doi.org/10.7554/eLife.81149

Cash S. A., M. D. Lorenzen, and F. Gould, (2019). 'The distribution and spread of naturally occurring Medea selfish genetic elements in the United States'. *Ecol. Evol.* 9: 1–10. https://doi.org/10.1002/ece3.5876

Champer J., R. Reeves, S. Y. Oh, et al., (2017). 'Novel CRISPR/Cas9 gene drive constructs reveal insights into mechanisms of resistance allele formation and drive efficiency in genetically diverse populations'. *PLoS Genet.* 13: 1–18. https://doi.org/10.1371/journal.pgen.1006796

Champer J., J. Liu, S. Y. Oh, et al., (2018). 'Reducing resistance allele formation in CRISPR gene drive'. *Proc. Natl. Acad. Sci.* 115: 5522–5527. https://doi.org/10.1073/pnas.1720354115

Charlat S., E. A. Hornett, J. H. Fullard, et al., (2007). 'Extraordinary flux in sex ratio'. *Science* 317: 214. https://doi.org/10.1126/science.1143369

Charlesworth, B. and Charlesworth, D., (1979). 'The maintenance and breakdown of distyly'. *Am. Nat.* 114: 499–513.

Charlwood J. D., (2020). *The Ecology of Malaria Vectors.* CRC Press, London.

Chen J.-M., N. Chuzhanova, P. D. Stenson, et al., (2005). 'Meta-analysis of gross insertions causing human genetic disease: novel mutational mechanisms and the role of replication slippage'. *Hum. Mutat.* 25: 207–221. https://doi.org/10.1002/humu.20133

Chesley P., and L. C. Dunn, (1936). 'The inheritance of taillessness (anury) in the house mouse'. *Genetics* 21: 525–536. https://doi.org/10.1093/genetics/22.2.331

Clarkson C., A. Miles, N. Harding, et al., (2021). 'The genetic architecture of target-site resistance to pyrethroid insecticides in the African malaria vectors *Anopheles gambiae* and *Anopheles coluzzii*'. *Mol. Ecol.* 30: 5303–5317. https://doi.org/10.1101/323980

Clutton-Brock T. H., P. N. M. Brotherton, A. F. Russell, et al., (2001). 'Cooperation, control, and concession in meerkat groups'. *Science* 291: 478–481. https://doi.org/10.1126/science.291.5503.478

Cohen D., (1985). 'The canine transmissible venereal tumor: a unique result of tumor progression'. *Adv. Cancer Res.* 43: 75–112. https://doi.org/10.1016/s0065-230x(08)60943-4

Colinvaux P., (1980). *Why Big Fierce Animals are Rare.* Allen & Unwin, London [Pelican Books, London].

Cook L. M., and I. J. Saccheri, (2013). 'The peppered moth and industrial melanism: evolution of a natural selection case study'. *Heredity* (Edinb). 110: 207–212. https://doi.org/10.1038/hdy.2012.92

Corliss J. B., J. Dymond, L. I. Gordon, et al., (1979). 'Submarine thermal springs on the Galapagos rift'. *Science* 203: 1073–1083. https://doi.org/10.1126/science.203.4385.1073

Corsi P., (2011). 'Lamarck, Jean-Baptiste: from myth to history', pp. 12–28 in *Transformations of Lamarckism: From Subtle Fluids to Molecular Biology*, edited by Jablonka E., Gissis S., MIT Press, Cambridge.

Cosmides L. M., and J. Tooby, (1981). 'Cytoplasmic inheritance and intragenomic conflict'. *J. Theor. Biol.* 89: 83–129. https://doi.org/10.1016/0022-5193(81)90181-8

Coyne J. A., and H. A. Orr, (2004). *Speciation*. Sinauer Associates, Sunderland.

Crespi B. J., (2000). 'The evolution of maladaptation'. *Heredity* (Edinb). 84: 623–629. https://doi.org/10.1046/j.1365-2540.2000.00746.x

Crews D., M. Grassman, and J. Lindzey, (1986). 'Behavioral facilitation of reproduction in sexual and unisexual whiptail lizards'. *Proc. Natl. Acad. Sci. USA.* 83: 9547–9550. https://doi.org/10.1073/pnas.83.24.9547

Crick F., (1981 [1982]). *Life Itself: Its Origin and Nature*. Simon & Schuster, New York.

Cronin H., (1991). *The Ant and the Peacock: Altruism and Sexual Selection from Darwin to Today*. Cambridge University Press, Cambridge.

Crow J. F., and M. Kimura, (1970 [2009]). *An Introduction to Population Genetics Theory*. Harper & Row, New York [Blackburn Press, Caldwell, NJ].

Crozier R. H., and Y. C. Crozier, (1993). 'The mitochondrial genome of the honeybee *Apis mellifera*: complete sequence and genome organization'. *Genetics* 133: 97–117. https://doi.org/10.1016/B978-0-12-813955-4.00026-X

Crozier R. H., and P. Pamilo, (1996). *Evolution of Social Insect Colonies: Sex Allocation and Kin Selection*. Oxford University Press, New York.

Curtis C. F., K. K. Grover, S. G. Suguna, et al., (1976). 'Comparative field cage tests of the population suppressing efficiency of three genetic control systems for *Aedes aegypti*'. *Heredity* (Edinb). 36: 11–29. https://doi.org/10.1038/hdy.1976.2

Darwin C., (1859). *On the Origin of Species by Means of Natural Selection, or the Preservation of Favoured Races in the Struggle for Life*. John Murray, London. http://darwin-online.org.uk/content/frameset?itemID=F373&viewtype=text&pageseq=1

Darwin C., (1868). *The Variation of Animals and Plants Under Domestication, Volume II*. John Murray, London. http://darwin-online.org.uk/content/frameset?itemID=F877.2&viewtype=text&pageseq=1

Darwin C., (1871). *The Descent of Man, Volume I, and Selection in Relation to Sex*. John Murray, London. http://darwin-online.org.uk/content/frameset?itemID=F937.1&viewtype=text&pageseq=1

Darwin C., (1887a). *The Life and Letters of Charles Darwin, including an Autobiographical Chapter. Volume III* (F. Darwin, Ed.). John Murray, London. http://darwin-online.org.uk/content/frameset?itemID=F1452.3&viewtype=text&pageseq=1

Darwin C., (1887b). *The Life and Letters of Charles Darwin, including an Autobiographical Chapter. Volume I* (F. Darwin, Ed.). John Murray, London. http://darwin-online.org.uk/content/frameset?viewtype=text&itemID=F1452.1&pageseq=1

Darwin C., (1887c). *The Life and Letters of Charles Darwin, including an Autobiographical Chapter. Volume II* (F. Darwin, ed.). John Murray, London. https://darwin-online.org.uk/content/frameset?itemID=F1452.2&viewtype=text&pageseq=1

Darwin C., (1958). *The Autobiography of Charles Darwin 1809–1882. With the original omissions restored. Edited and with appendix and notes by his grand-daughter Nora Barlow* (N. Barlow, Ed.). Collins, London. http://darwin-online.org.uk/content/frameset?itemID=F1497&viewtype=text&pageseq=1

Dawkins R., (1976 [2006]). *The Selfish Gene*. Oxford University Press, New York.

Dawkins R., (1982 [1983]). *The Extended Phenotype: The Gene as the Unit of Selection*. W. H. Freeman & Co. [Oxford University Press, New York].

Dawkins R., (1990). 'Parasites, desiderata lists and the paradox of the organism'. *Parasitology* 100: S63–S73. https://doi.org/10.1017/S0031182000073029

Dawkins R., (1998). *Unweaving the Rainbow: Science, Delusion and the Appetite for Wonder*. Houghton Mifflin, Boston.

Dawkins R., and Y. Wong, (2004 [2005]). *The Ancestor's Tale: A Pilgrimage to the Dawn of Life*. Weidenfeld & Nicolson, London [Phoenix, London].

Dawkins R., (2006 [2007]). *The God Delusion*. Bantham Press, London [Transworld Publishers, London].

Dawkins R., (2013 [2014]). *An Appetite for Wonder: The Making of a Scientist*. Bantham Press, London [Transworld Publishers, London].

Dawkins R., and C. J. Venter, (2016). 'The Gene-Centric View: a conversation', pp. 189–212 in *Life: The Leading Edge of Evolutionary Biology, Genetics, Anthropology, and Environmental Science*, edited by Brockman J., HarperCollins, New York.

De Duve C., (1991). *Blueprint for a Cell: The Nature and Origin of Life*. Portland Press, London.

Deininger P. L., and M. A. Batzer, (2002). 'Mammalian retroelements'. *Genome Res.* 12: 1455–1465. https://doi.org/10.1101/gr.282402

Deredec A., A. Burt, and H. C. J. Godfray, (2008). 'The population genetics of using homing endonuclease genes in vector and pest management'. *Genetics* 179: 2013–2026. https://doi.org/10.1534/genetics.108.089037

Desmond A., and J. Moore, (1991). *Darwin*. Michael Joseph, London.

Diamond J., (1991 [1992]). *The Rise and Fall of the Third Chimpanzee: How Our Animal Heritage Affects the Way We Live*. Radius, London [Vintage, London].

Dobzhansky T., (1937 [1982]). *Genetics and the Origin of Species*. Columbia University Press, New York.

Drummond H., (1894). *The Ascent of Man*. Hodder and Stoughton, London.

Dunham I., A. Kundaje, S. F. Aldred, et al., (2012). 'An integrated encyclopedia of DNA elements in the human genome'. *Nature* 489: 57–74. https://doi.org/10.1038/nature11247

Dyson F. J., (1982). 'A model for the origin of life'. *J. Mol. Evol.* 18: 344–350. https://doi.org/10.1007/BF01733901

Dyson F. J., (1984 [1999]). *Origins of Life*. Cambridge University Press, New York.

Eigen M., (1971). 'Self-organization of matter and the evolution of biological macromolecules'. *Naturwissenschaften* 58: 465–523. https://doi.org/10.1007/BF00623322

Eigen M., (1992 [1996]). *Steps Towards Life: A Perspective on Evolution*. Oxford University Press, New York.

Ezkurdia I., D. Juan, J. M. Rodriguez, et al., (2014). 'Multiple evidence strands suggest that theremay be as few as 19000 human protein-coding genes'. *Hum. Mol. Genet.* 23: 5866–5878. https://doi.org/10.1093/hmg/ddu309

Fisher R. A., (1914). 'Some hopes of a eugenist'. *Eugen. Rev.* 5: 309–315.

Fisher R. A., (1930 [2011]). *The Genetical Theory of Natural Selection.* Clarendon Press, Oxford [Oxford University Press, New York].

Foster K. R., F. L. W. Ratnieks, and T. Wenseleers, (2000). 'Spite in social insects'. *Trends Ecol. Evol.* 15: 469–470. https://doi.org/10.1016/S0169-5347(00)01978-9

Foster K. R., T. Wenseleers, and F. L. W. Ratnieks, (2001). 'Spite: Hamilton's unproven theory'. *Ann. Zool. Fennici* 38: 229–238.

Foster K. R., (2011). 'The sociobiology of molecular systems'. *Nat. Rev. Genet.* 12: 193–203. https://doi.org/10.1038/nrg2903

Fox L. R., (1975). 'Cannibalism in natural populations'. *Annu. Rev. Ecol. Syst.* 6: 87–106. https://doi.org/10.1146/annurev.es.06.110175.000511

Fox S. W., (1980). 'Life from an orderly cosmos'. *Naturwissenschaften* 67: 576–581. https://doi.org/10.1007/BF00396536

Frank S. A., (1998). *Foundations of Social Evolution.* Princeton University Press, Princeton, NJ.

Frank S. A., (2003). 'Repression of competition and the evolution of cooperation'. *Evolution* (NY) 57: 693–705. https://doi.org/10.1111/j.0014-3820.2003.tb00283.x

Fry I., (1999). *The Emergence of Life on Earth: A Historical and Scientific Overview.* Free Association Books, London.

Gadagkar R., (1993). 'Can animals be spiteful?' *Trends Ecol. Evol.* 8: 232–234. https://doi.org/10.1016/0169-5347(93)90196-V

Galliard J. F. Le, P. S. Fitze, R. Ferrière, and J. Clobert, (2005). 'Sex ratio bias, male aggression, and population collapse in lizards'. *Proc. Natl. Acad. Sci. USA.* 102: 18231–18236. https://doi.org/10.1073/pnas.0505172102.

Galton F., (1869). *Hereditary Genius: An Enquiry into Its Law and Consequences.* Macmillan and Co., London.

Gardner A., and S. A. West, (2004). 'Spite and the scale of competition'. *J. Evol. Biol.* 17: 1195–1203. https://doi.org/10.1111/j.1420-9101.2004.00775.x.

Gardner A., and S. A. West, (2010) 'Greenbeards'. *Evolution* (NY) 64: 25–38. https://doi.org/10.1111/j.1558-5646.2009.00842.x.

Gardner A., S. A. West, and A. Buckling, (2004). 'Bacteriocins, spite and virulence'. *Proc. R. Soc. London B* 271: 1529–1535. https://doi.org/10.1098/rspb.2004.2756.

Gardner A., I. C. W. Hardy, P. D. Taylor, et al., (2007). 'Spiteful soldiers and sex ratio conflict in polyembryonic parasitoid wasps'. *Am. Nat.* 169: 519–533. https://doi.org/10.1086/512107.

Gardner A., (2009). 'Adaptation as organism design'. *Biol. Lett.* 5: 861–864. https://doi.org/10.1098/rsbl.2009.0674.

Gardner A., (2013). 'Adaptation of individuals and groups', pp. 99–116 in *From Groups to Individuals: Evolution and Emerging Individuality*, edited by Bouchard F., Huneman P., MIT Press, Cambridge.

Gardner A., and F. Úbeda, (2017). 'The meaning of intragenomic conflict'. *Nat. Ecol. Evol.* 1: 1807–1815. https://doi.org/10.1038/s41559-017-0354-9

Gartler S. M., (1955). 'The evolutionary problem of genetic disease'. *Eugen. Q.* 2: 40–45. https://doi.org/10.1080/19485565.1955.9987219

Gershenson S., (1928). 'A new sex-ratio abnormality in *Drosophila obscura*'. *Genetics* 13: 488–507. https://doi.org/10.1093/genetics/14.1.127

Ghiselin M. T., (1974). *The Economy of Nature and the Evolution of Sex*. University of California Press, Berkeley and Los Angeles.

Gillespie, J. H. (2000). 'Genetic drift in an infinite population: the pseudohitchhiking model'. *Genetics* 155: 909–919.

Giron D., D. D. Dunn, I. C. W. Hardy, et al., (2004). 'Aggression by polyembryonic wasp soldiers correlates with kinship but not resource competition'. *Nature* 430: 676–679. https://doi.org/10.1038/nature02631.1.

Giron D., K. G. Ross, and M. R. Strand, (2007). 'Presence of soldier larvae determines the outcome of competition in a polyembryonic wasp'. *J. Evol. Biol.* 20: 165–172. https://doi.org/10.1111/j.1420-9101.2006.01212.x

Godfrey-Smith P., (2009 [2015]). *Darwinian Populations and Natural Selection*. Oxford University Press, New York.

Goldschmidt R., (1929). 'Experimentelle mutation und das problem der sogenannten parallelinduktion. Versuche an Drosophila'. *Biol. Zent. Bl.* 49: 437–448.

Goodier J. L., and H. H. Kazazian, (2008). 'Retrotransposons revisited: the restraint and rehabilitation of parasites'. *Cell* 135: 23–35. https://doi.org/10.1016/j.cell.2008.09.022

Gould S. J., and R. C. Lewontin, (1979). 'The spandrels of San Marco and the Panglossian paradigm: a critique of the adaptationist programme'. *Proc. R. Soc. B* 205: 581–598.

Gould S. J., (1989 [1990]). *Wonderful Life: The Burgess Shale and the Nature of History*. W. W. Norton & Company, New York.

Gould S. J., (1996 [1997]). *Life's Grandeur: The Spread of Excellence from Plato to Darwin*. Jonathan Cape, London [Random House, London].

Gould S. J., (1999 [2002]). *Rocks of Ages: Science and Religion in the Fullness of Life*. Ballantine Books, New York [Random House, London].

Gould S. J., (2002). *The Structure of Evolutionary Theory*. Harvard University Press, Cambridge, MA.

Grafen A., (1984). 'Natural selection, kin selection and group selection', pp. 62–84 in *Behavioural Ecology*, edited by Krebs J. R., Davies N. B., Blackwell Science, Oxford.

Grafen A., (1985). 'A geometric view of relatedness'. *Oxford Surv. Evol. Biol.* 2: 28–89.

Grafen A., (1998). 'Green beard as death warrant'. *Nature* 394: 521–523. https://doi.org/10.1038/28948

Grafen A., (2006a). 'The intellectual contribution of *The Selfish Gene* to evolutionary theory', pp. 66–74 in *Richard Dawkins: How a Scientist Changed the Way We Think*, edited by Grafen A., Ridley M., Oxford University Press, New York.

Grafen A., (2006b). 'Optimization of inclusive fitness'. *J. Theor. Biol.* 238: 541–563. https://doi.org/10.1016/j.jtbi.2005.06.009

Griffin A. S., S. A. West, and A. Buckling, (2004). 'Cooperation and competition in pathogenic bacteria'. *Nature* 430: 1024–1027. https://doi.org/10.1038/nature02802.1.

Grosberg R. K., and R. R. Strathmann, (2007). 'The evolution of multicellularity: a minor major transition?' *Annu. Rev. Ecol. Evol. Syst.* 38: 621–654. https://doi.org/10.1146/annurev.ecolsys.36.102403.114735

Gruenheit N., K. Parkinson, B. Stewart, et al., (2017) 'A polychromatic "greenbeard" locus determines patterns of cooperation in a social amoeba'. *Nat. Commun.* 14171: 1–9. https://doi.org/10.1038/ncomms14171

Gummere G. R., P. J. McCormick, and D. Bennett, (1986). 'The influence of genetic background and the homologous chromosome 17 on t-haplotype transmission ratio distortion in mice'. *Genetics* 114: 235–245. https://doi.org/10.1093/genetics/114.1.235

Gyllenberg M., K. Parvinen, and U. Dieckmann, (2002). 'Evolutionary suicide and evolution of dispersal in structured metapopulations'. *J. Math. Biol.* 45: 79–105. https://doi.org/10.1007/s002850200151

Haig D., and A. Grafen, (1991). 'Genetic scrambling as a defence against meiotic drive'. *J. Theor. Biol.* 153: 531–558. https://doi.org/10.1016/S0022-5193(05)80155-9

Haig D., (1996). 'Gestational drive and the green-bearded placenta'. *Proc. Natl. Acad. Sci.* 93: 6547–6551. https://doi.org/10.1073/pnas.93.13.6547

Haig D., (1997). 'The social gene', pp. 284–304 in Behavioural Ecology: An Evolutionary Approach, edited by Krebs J. R., Davies N. B., Blackwell Science, Oxford.

Haig D., (2010). 'The huddler's dilemma: a cold shoulder or a warm inner glow', pp. 107–109 in *Social Behaviour: Genes, Ecology and Evolution*, edited by Székely T., Moore A. J., Komdeur J., Oxford University Press.

Haig D., (2012). 'The strategic gene'. *Biol. Philos.* 27: 461–479. https://doi.org/10.1007/s10539-012-9315-5

Haig, D., (2019). 'Cooperation and conflict in human pregnancy'. *Curr. Biol.* 29: R455–R458.

Haldane J. B. S., (1932). *The Causes of Evolution*. Longmans, Green and Co., London.

Haldane J. B. S., (1955). 'Population genetics'. *New Biol.* 18: 34–51.

Haldane J. B. S., (1929 [1968]). *Science and Life: Essays of a Rationalist* (J. Maynard Smith, Ed.). Pemberton Publishing, London.

Haldane J. B. S., (1985). *On Being the Right Size: And other Essays* (J. Maynard Smith, Ed.). Oxford University Press, Oxford.

Hamede R. K., H. Mccallum, and M. Jones, (2008). 'Seasonal, demographic and density-related patterns of contact between Tasmanian devils (*Sarcophilus harrisii*): implications for transmission of devil facial tumour disease'. *Austral Ecol.* 33: 614–622. https://doi.org/10.1111/j.1442-9993.2007.01827.x

Hamilton W. D., (1963). 'The evolution of altruistic behavior'. *Am. Nat.* 97: 354–356. https://doi.org/10.1086/497114

Hamilton W. D., (1964a). 'The genetical evolution of social behaviour II'. *J. Theor. Biol.* 7: 17–52. https://doi.org/10.1016/0022-5193(64)90039-6

Hamilton W. D., (1964b). 'The genetical evolution of social behaviour I'. *J. Theor. Biol.* 7: 1–16. https://doi.org/10.1016/0022-5193(64)90038-4

Hamilton W. D., (1967). 'Extraordinary sex ratios'. *Science* 156: 477–488. https://doi.org/10.1126/science.156.3774.477

Hamilton W. D., (1970). 'Selfish and spiteful behaviour in an evolutionary model'. *Nature* 228: 1218–1220. https://doi.org/10.1038/2281218a0

Hamilton W. D., (1971). 'Selection of selfish and altruistic behavior in some extreme models', pp. 57–91 in *Man and Beast: Comparative Social Behavior*, edited by Eisenberg J. F., Dillon W. S. Smithsonian Institution Press, Washington.

Hamilton W. D., (1972). 'Altruism and related phenomena, mainly in social insects'. *Annu. Rev. Ecol. Syst.* 3: 193–232. https://doi.org/10.1146/annurev.es.03.110172.001205

Hamilton W. D., (1975). 'Innate Social Aptitudes of Man: an approach from evolutionary genetics', pp. 133–153 in *ASA Studies 4: Biosocial Anthropology*, edited by Fox R., Malaby Press, London.

Hamilton W. D., (1987). 'Discriminating nepotism: expectable, common, overlooked', pp. 417–437 in *Kin Recognition in Animals*, edited by Fletcher D. J. C., Michener C. D., Wiley, New York.

Hamilton W. D., (2001). *Narrow Roads of Gene Land: Volume 2*. Oxford University Press, New York.

Hanahan D., and R. Weinberg, (2011). 'Hallmarks of cancer: the next generation'. *Cell* 144: 646–674. https://doi.org/10.1016/j.cell.2011.02.013

Hardin G., (1968). 'The tragedy of the commons'. *Science* 162: 1243–1248. https://doi.org/10.1126/science.162.3859.1243

Harris S., (2010 [2012]). *The Moral Landscape: How Science can Determine Human Values*. Batham Press, London [Transworld Publishers, London].

Hastings I. M., (1999). 'The costs of sex due to deleterious intracellular parasites'. *J. Evol. Biol.* 12: 177–183. https://doi.org/10.1046/j.1420-9101.1999.00019.x

Hatchwell B. J., (2009). 'The evolution of cooperative breeding in birds: kinship, dispersal and life history'. *Philos. Trans. R. Soc. Lond. B. Biol. Sci.* 364: 3217–3227. https://doi.org/10.1098/rstb.2009.0109

Havird J. C., E. S. Forsythe, A. M. Williams, et al., (2019). 'Selfish mitonuclear conflict'. *Curr. Biol.* 29: R496–R511. https://doi.org/10.1016/j.cub.2019.03.020

Hendry A. P., (2016 [2020]). *Eco-Evolutionary Dynamics*. Princeton University Press, Princeton, NJ.

Henikoff S., K. Ahmad, and H. S. Malik, (2001). 'The centromere paradox: stable inheritance with rapidly evolving DNA'. *Science* 293: 1098–1102. https://doi.org/10.1126/science.1062939

Herrmann B. G., B. Koschorz, K. Wert, et al., (1999). 'A protein kinase encoded by the t complex responder gene causes non-mendelian inheritance'. *Nature* 402: 141–146. https://doi.org/10.1038/45970

Hickey W. A., and G. B. Craig, (1966a). 'Genetic distortion of sex ratio in a mosquito, *Aedes aegypti*'. *Genetics* 53: 1177–1196. https://doi.org/10.1093/genetics/53.6.1177

Hickey W. A., and G. B. Craig, (1966b). 'Distortion of sex ratio in populations of *Aedes aegypti*'. *Can. J. Genet. Cytol.* 8: 260–278. https://doi.org/10.1139/g66-033

Hikosaka K., K. Kita, and K. Tanabe, (2013). 'Diversity of mitochondrial genome structure in the phylum Apicomplexa'. *Mol. Biochem. Parasitol.* 188: 26–33. https://doi.org/10.1016/j.molbiopara.2013.02.006

Hilgenboecker K., P. Hammerstein, P. Schlattmann, et al., (2008). 'How many species are infected with Wolbachia?—a statistical analysis of current data'. *FEMS Microbiol. Lett.* 281: 215–220. https://doi.org/10.1111/j.1574-6968.2008.01110.x

Holland B., and W. R. Rice, (1999). 'Experimental removal of sexual selection reverses intersexual antagonistic coevolution and removes a reproductive load'. *Proc. Natl. Acad. Sci. USA.* 96: 5083–5088. https://doi.org/10.1073/pnas.96.9.5083

Hornett E. A., S. Charlat, N. Wedell, et al., (2009). 'Rapidly shifting sex ratio across a species range'. *Curr. Biol.* 19: 1628–1631. https://doi.org/10.1016/j.cub.2009.07.071

Horowitz N. H., (1959). 'On defining life', pp. 106–107 in *The Origin of Life on the Earth*, edited by Clark F., Synge R. L. M., Pergamon Press, London.

Hotzy C., and G. Arnqvist, (2009). 'Sperm competition favors harmful males in seed beetles'. *Curr. Biol.* 19: 404–407. https://doi.org/10.1016/j.cub.2009.01.045

Hu W., Z.-D. Jiang, F. Suo, et al., (2017). 'A large gene family in fission yeast encodes spore killers that subvert Mendel's law'. *Elife* 6: 1–19. https://doi.org/10.7554/elife.26057

Hume D., (1739). *A Treatise on Human Nature*. John Noon, London. https://davidhume.org/texts/t/full

Hume, D. (1779). *Dialogues Concerning Natural Religion*. William Strahan, London. https://davidhume.org/texts/d/full

Hurst L. D., (1991). 'The evolution of cytoplasmic incompatibility or when spite can be successful'. *J. Theor. Biol.* 148: 269–277. https://doi.org/10.1016/S0022-5193(05)80344-3

Hurst L. D., and A. Pomiankowski, (1991). 'Maintaining Mendelism: might prevention be better than cure?' *BioEssays* 13: 489–490. https://doi.org/10.1001/jama.295.20.2349

Hurst L. D., A. Atlan, and B. O. Bentsson, (1996). 'Genetic conflicts'. *Q. Rev. Biol.* 71: 317–364. https://doi.org/10.1086/419442

Hurst G. D. D., and G. A. T. McVean, (1998). 'Selfish genes in a social insect'. *Trends Ecol. Evol.* 13: 434–435. https://doi.org/10.1016/S0169-5347(98)01479-7

Hurst L. D., and J. P. Randerson, (2000). 'Transitions in the evolution of meiosis'. *J. Evol. Biol.* 13: 466–479. https://doi.org/10.1046/j.1420-9101.2000.00182.x

Hurst G. D. D., and J. H. Werren, (2001). 'The role of selfish genetic elements in eukaryotic evolution'. *Nat. Rev. Genet.* 2: 597–606. https://doi.org/10.1038/35084545

Hurst L. D., (2022). 'Selfish centromeres and the wastefulness of human reproduction'. *PLoS Biol.* 20: e3001671. https://doi.org/10.1371/journal.pbio.3001671

Huxley J. S., (1912 [2011]). *The Individual in the Animal Kingdom*. Cambridge University Press, Cambridge.

Huxley J., (1942). *Evolution: The Modern Synthesis*. George Allen & Unwin, London.

Inglis R. F., A. Gardner, P. Cornelis, and A. Buckling, (2009). 'Spite and virulence in the bacterium *Pseudomonas aeruginosa*'. *Proc. Natl. Acad. Sci. USA.* 106: 5703–5707. https://doi.org/10.1073/pnas.0810850106

Inglis R. F., P. G. Roberts, A. Gardner, et al., (2011). 'Spite and the scale of competition in *Pseudomonas aeruginosa*'. *Am. Nat.* 178: 276–285. https://doi.org/10.1086/660827

Ingram V. M., (2004). 'Sickle-cell anemia hemoglobin: the molecular biology of the first "molecular disease"—the crucial importance of serendipity'. *Genetics* 167: 1–7. https://doi.org/10.1534/genetics.167.1.1

Jablonski D., (2008) 'Species selection: theory and data'. *Annu. Rev. Ecol. Evol. Syst.* 39: 501–524. https://doi.org/10.1146/annurev.ecolsys.39.110707.173510

Jacob F., (1977). 'Evolution and tinkering'. *Science* 196: 1161–1166. https://doi.org/10.1126/science.860134

Jasmin J. N., and C. Zeyl, (2014). 'Rapid evolution of cheating mitochondrial genomes in small yeast populations'. *Evolution* (NY) 68: 269–275. https://doi.org/10.1111/evo.12228

Jiggins F. M., J. P. Randerson, G. D. D. Hurst, et al., (2002). 'How can sex ratio distorters reach extreme prevalences? Male-killing Wolbachia are not suppressed and have near-perfect vertical transmission efficiency in *Acraea encedon*'. *Evolution* (NY) 56: 2290–2295. https://doi.org/10.1111/j.0014-3820.2002.tb00152.x

John U., Y. Lu, S. Wohlrab, et al., (2019). 'An aerobic eukaryotic parasite with functional mitochondria that likely lacks a mitochondrial genome'. *Sci. Adv.* 5: 1–12. https://doi.org/10.1126/sciadv.aav1110

Judson H., (1979 [1995]). *The Eighth Day of Creation*. Jonathan Cape, London [Penguin Books, London].

Kauffman S., (1993). *The Origins of Order*. Oxford University Press, New York.

Keller L., M. Milinski, M. Frischknecht, et al., (1993). 'Spiteful animals still to be discovered'. *Trends Ecol. Evol.* 9: 228. https://doi.org/10.1016/0169-5347(94)90205-4

Keller L., and K. G. Ross, (1998) 'Selfish genes: a green beard in the red fire ant'. *Nature* 394: 573–575. https://doi.org/10.1038/29064

Kettlewell H. B. D., (1958). 'A survey of the frequencies of *Biston betularia* (L.) (Lep.) and its melanic forms in Great Britain'. *Heredity* (Edinb.) 12: 51–72. https://doi.org/10.1038/hdy.1958.4

Keynes, J. M., (1946). 'Newton, the Man', pp. 27–34 in *Newton Tercentenary Celebrations, 15–19 July 1946*. Cambridge University Press, Cambridge. https://mathshistory.st-andrews.ac.uk/Extras/Keynes_Newton/

Kitazaki K., and T. Kubo, (2010). 'Cost of having the largest mitochondrial genome: evolutionary mechanism of plant mitochondrial genome'. *J. Bot.* (2010): 1–12. https://doi.org/10.1155/2010/620137

Klein J., P. Sipos, and F. Figueroa, (1984). 'Polymorphism of t-complex genes in European wild mice'. *Genet. Res.* 44: 39–46. https://doi.org/10.1017/S0016672300026239

Knoll A. H., (2003 [2005]). *Life on a Young Planet: The First Three Billion Years of Evolution*. Princeton University Press, Woodstock, UK.

Knowlton N., and G. A. Parker, (1979). 'An evolutionarily stable strategy approach to indiscriminate spite'. *Nature* 279: 419–421. https://doi.org/10.1038/279419a0

Kohn M., (2004). *A Reason for Everything: Natural Selection and the English Imagination*. Faber & Faber, London.

Kropotkin P. A., (1902). *Mutual Aid: A Factor of Evolution*. McClure Phillips & Co., New York. https://archive.org/details/mutualaidfactoroookrop_1/page/n3/mode/2up

Krupp D. B., (2013). 'How to distinguish altruism from spite (and why we should bother)'. *J. Evol. Biol.* 26: 2746–2749. https://doi.org/10.1111/jeb.12253

Kuhn T. S., (1962). *The Structure of Scientific Revolutions*. University of Chicago Press, Chicago.

Laberge A. M., M. Jomphe, L. Houde, et al., (2005). 'A "Fille du Roy" introduced the T14484C leber hereditary optic neuropathy mutation in French Canadians'. *Am. J. Hum. Genet.* 77: 313–317. https://doi.org/10.1086/432491

Lamarck J. B., (1809 [1963]). *Zoological Philosophy: An Exposition with Regard to the Natural History of Animals*. Musée d'Histoire Naturelle, Paris [Hafner Publishing Co., New York and London].

Leeks A., S. A. West, and M. Ghoul, (2021). 'The evolution of cheating in viruses'. *Nat. Commun.* 12: 1–14. https://doi.org/10.1038/s41467-021-27293-6

Lehmann L., K. Bargum, and M. Reuter, (2006). 'An evolutionary analysis of the relationship between spite and altruism'. *J. Evol. Biol.* 19: 1507–1516. https://doi.org/10.1111/j.1420-9101.2006.01128.x

Leigh E. G., (1971). *Adaptation and Diversity: Natural History and the Mathematics of Evolution*. Freeman, Cooper & Company, New York.

Leigh E. G., (1977). 'How does selection reconcile individual advantage with the good of the group?' *Proc. Natl. Acad. Sci.* 74: 4542–4546. https://doi.org/10.1073/pnas.74.10.4542

Levin S. R., and A. Grafen, (2019). 'Inclusive fitness is an indispensable approximation for understanding organismal design'. *Evolution* (NY) evo.13739. https://doi.org/10.1111/evo.13739

Lewontin R. C., (1970). 'The units of selection'. *Annu. Rev. Ecol. Evol. Syst.* 1: 1–18. https://doi.org/10.1146/annurev.es.01.110170.000245

López Hernández J. F., and S. E. Zanders, (2018). 'Veni, vidi, vici: the success of wtf meiotic drivers in fission yeast'. *Yeast* 35: 447–453. https://doi.org/10.1002/yea.3305

Lyttle T. W., (1977). 'Experimental population genetics of meiotic drive systems. I. Pseudo Y chromosomal drive as a means of eliminating cage populations of *Drosophila melanogaster*'. *Genetics* 86: 413–445. https://doi.org/10.1093/genetics/86.2.413

Lyttle T. W., (1979). 'Experimental population genetics of meiotic drive systems. II. Accumulation of genetic modifiers of segregation distorter (SD) in laboratory populations'. *Genetics* 91: 339–357. https://doi.org/10.1093/genetics/91.2.339

Lyttle T. W., (1981). 'Experimental population genetics of meiotic drive systems. III. Neutralisation of sex-ratio distortion in Drosophila through sex-chromosome aneuploidy'. *Genetics* 98: 317–334. https://doi.org/10.1093/genetics/98.2.317

MacArthur R., and E. O. Wilson, (1967 [2001]). *The Theory of Island Biogeography*. Princeton University Press, Princeton, NJ.

Madgwick P. G., B. Stewart, L. J. Belcher, et al., (2018) 'Strategic investment explains patterns of cooperation and cheating in a microbe'. *Proc. Natl. Acad. Sci.* 115: E4823–E4832. https://doi.org/10.1073/pnas.1716087115

Madgwick P. G., L. J. Belcher, and J. B. Wolf, (2019). 'Greenbeard genes: theory and reality'. *Trends Ecol. Evol.* 34: 1092–1103. https://doi.org/10.1016/j.tree.2019.08.001

Madgwick P. G., (2020). 'Spite and the geometry of negative relatedness'. *Am. Nat.* 196: 1–8. https://doi.org/10.1086/710764

Madgwick P., (2021). 'Agency in evolutionary biology', pp. 211–230 in *Karl Popper's Science and Philosophy*, edited by Parusniková Z., Merritt D., Springer, Cham.

Madgwick P. G., and R. Kanitz, (2021). 'Evolution of resistance under alternative models of selective interference'. *J. Evol. Biol.* 34: 1–16. https://doi.org/10.1111/jeb.13919

Madgwick P. G., and J. B. Wolf, (2021). 'Evolutionary robustness of killer meiotic drives'. *Evol. Lett.* 5: 1–10. https://doi.org/10.1002/evl3.255

Madgwick P. G., T. Tunstall, and R. Kanitz, (in press). 'Evolutionary rescue in resistance to pesticides'.

Majerus M. E. N., (1998 [2005]). *Melanism: Evolution in Action*. Oxford University Press, New York.

Malik H. S., and S. Henikoff, (2002). 'Conflict begets complexity: the evolution of centromeres'. *Curr. Opin. Genet. Dev.* 12: 711–718. https://doi.org/10.1016/S0959-437X(02)00351-9

Malthus T. R., (1798 [2015]). *An Essay on the Principle of Population and Other Writings*. J. Johnson, London [Penguin Books, London].

Mangel M., and F. J. Samaniego, (1984). 'Abraham Wald's work on aircraft survivability'. *J. Am. Stat. Assoc.* 79: 256–267. https://doi.org/10.2307/2288257

Margulis L., (1970). *Origin of Eukaryotic Cells: Evidence and Research Implications for a Theory of the Origin and Evolution of Microbial, Plant, and Animal Cells on the Precambrian Earth*. Yale University Press, New Haven, CT.

Marshall J. A. R., (2015). *Social Evolution and Inclusive Fitness Theory*. Princeton University Press, Princeton, NJ.

Martin S. J., M. Beekman, T. C. Wossler, et al., (2002a). 'Parasitic Cape honeybee workers, Apis mellifera capensis, evade policing'. Nature 415: 163–165. https://doi.org/10.1038/415163a

Martin S., T. Wossler, and P. Kryger, (2002b). 'Usurpation of African Apis mellifera scutellata colonies by parasitic *Apis mellifera capensis* workers'. *Apidologie* 33: 215–232. https://doi.org/10.1051/apido

May R. M., (1976). 'Simple mathematical models with very complicated dynamics'. *Nature* 261: 459–467. https://doi.org/10.1038/261459a0

Maynard Smith J., (1958 [1993]). *The Theory of Evolution*. Penguin Books, London [Cambridge University Press, Cambridge].

Maynard Smith J., (1964). 'Group selection and kin selection'. *Nature* 201: 1145–1147. https://doi.org/10.1038/2011145a0

Maynard Smith J., (1976). 'Group selection'. *Q. Rev. Biol.* 51: 277–283. https://doi.org/10.1086/409311

Maynard Smith J., (1988). 'Evolutionary progress and the levels of selection', pp. 219–230 in *Evolutionary Progress*, edited by Nitecki M., University of Chicago Press, Chicago.

Maynard Smith J., (1989). 'The causes of extinction'. *Philos. Trans. R. Soc. London, B* 325: 241–252. https://doi.org/10.1098/rstb.1989.0086

Maynard Smith J., and J. Haigh, (1974). 'The hitch-hiking effect of a favourable gene'. *Genet. Res.* 23: 23–35. https://doi.org/10.1017/S0016672308009579

Maynard Smith J., and E. Szathmáry, (1995 [2010]). *The Major Transitions in Evolution*. Oxford University Press, New York.

Mayr E., (1982). *The Growth of Biological Thought: Diversity, Evolution, and Inheritance*. Harvard University Press, Cambridge, MA.

McCallum H., D. M. Tompkins, M. Jones, et al., (2007). 'Distribution and impacts of Tasmanian devil facial tumor disease'. *Ecohealth* 4: 318–325. https://doi.org/10.1007/s10393-007-0118-0

McCauley, D. E. and Taylor, D. R., (1997). 'Local population structure and sex ratio: evolution in gynodioecious plants'. *Am. Nat.* 150: 406–419.

McGrath A. E., (2008 [2011]). *Darwinism and the Divine: Evolutionary Thought and Natural Theology*. Wiley-Blackwell, Oxford.

Medawar P., (1967 [1969]). *The Art of the Soluble*. Methuen, London [Pelican Books, London].

Mendel G., (1866). 'Versuche über Pflanzenhybriden'. *Verhandlungen des natur-forschenden Vereines Brünn* 4: 3–47.

Merwe P. L. Van Der, H. W. Weymar, and N. N. Kalis, (1994). 'The extent of progressive familial heart block type I—a new perspective'. *Cardiovasc. J. South. Africa* 5: 125–126.

Metzger M. J., C. Reinisch, J. Sherry, et al., (2015). 'Horizontal transmission of clonal cancer cells causes leukemia in soft-shell clams'. *Cell* 161: 255–263. https://doi.org/10.1016/j.cell.2015.02.042

Michod R. E., (1995). *Eros and Evolution: A Natural Philosophy of Sex*. Addison-Wesley, New York.

Michod R. E., (1999 [2000]). *Darwinian Dynamics: Evolutionary Transitions in Fitness and Individuality*. Princeton University Press, Princeton, NJ.

Mill J. S., (1863). *Utilitarianism*. Parker, Son, and Bourn, London. https://archive.org/details/a592840000milluoft/page/n3/mode/2up

Miller S. L., (1953). 'A production of amino acids under possible primitive earth conditions'. *Science* 117: 528–529. https://doi.org/10.1126/science.117.3046.528

Milot E., C. Moreau, A. Gagnon, et al., (2017). 'Mother's curse neutralizes natural selection against a human genetic disease over three centuries'. *Nat. Ecol. Evol.* 1: 1400–1406. https://doi.org/10.1038/s41559-017-0276-6

Miné M., J. Chen, I. Desguerre, et al., (2006). 'A large genomic deletion in the PDHX gene caused by the retrotranspositional insertion of a full-length LINE-1 element'. *Hum. Mutat.* 28: 137–142. https://doi.org/10.1002/humu

Monod J., (1970 [1972]). *Chance and Necessity: An Essay on the Natural Philosophy of Modern Biology*. Seuil, Paris [Collins, London].

Morandini V., and M. Ferrer, (2015). 'Sibling aggression and brood reduction: A review'. *Ethol. Ecol. Evol.* 27: 2–16. https://doi.org/10.1080/03949370.2014.880161

Morgan T. H., A. H. Sturtevant, H. J. Muller, et al., (1915). *The Mechanism of Mendelian Heredity*. Henry Holt and Company, New York. http://www.esp.org/books/morgan/mechanism/facsimile/

Mumme R. L., (1992). 'Do helpers increase reproductive success? An experimental analysis in the Florida scrub jay'. *Behav. Ecol. Sociobiol.* 31: 319–328. https://doi.org/10.1007/BF00177772

Murgia C., J. K. Pritchard, S. Y. Kim, Aet al., (2008). 'Clonal origin and evolution of a transmissible cancer'. *Cell* 126: 477–487. https://doi.org/10.1016/j.cell.2006.05.051.Clonal

NCARDRS, (2019). NCARDRS Statistics 2019. https://www.gov.uk/government/publications/ncardrs-congenital-anomaly-annual-data/ncardrs-statistics-2019-summary-report#prevalence

Nesse R. M., (2005). 'Maladaptation and natural selection'. *Q. Rev. Biol.* 80: 62–70. https://doi.org/10.1086/431026

Niehus R., N. M. Oliveira, A. G. Fletcher, et al., (2021). 'The evolution of strategy in bacterial warfare'. *ELife* 1–34. https://doi.org/10.7554/eLife.69756

Nuckolls N. L., M. A. B. Núñez, M. T. Eickbush, et al., (2017). 'Wtf Genes Are Prolific Dual Poison–Antidote Meiotic Drivers'. *Elife* 6: 1–22. https://doi.org/10.7554/eLife.26033

Ogunlade S. T., M. T. Meehan, A. I. Adekunle, et al., (2021). 'A review: Aedes-borne arboviral infections, controls and wolbachia-based strategies'. *Vaccines* 9: 1–23. https://doi.org/10.3390/vaccines9010032

Ogunlade S. T., A. I. Adekunle, M. T. Meehan, et al., (2023). 'Quantifying the impact of Wolbachia releases on dengue infection in Townsville, Australia'. *Sci. Rep.* 13: 1–12. https://doi.org/10.1038/s41598-023-42336-2

ONS, (2023a). Housing affordability in England and Wales: (2022). https://www.ons.gov.uk/peoplepopulationandcommunity/housing/bulletins/housingaffordabilityinenglandandwales/2021

ONS, (2023b). Birth characteristics in England and Wales: 2021. https://www.ons.gov.uk/peoplepopulationandcommunity/birthsdeathsandmarriages/livebirths/bulletins/birthcharacteristicsinenglandandwales/2021#:~:text=5.-,Ethnicity,by ethnicity of the baby.

Oparin A. I., (1967). 'The origin of life', pp. 199–234 in *The Origin of Life*, edited by Bernal J. D., Weidenfeld & Nicolson, London.

Orgel L. E., and S. L. Miller, (1974). *The Origins of Life on the Earth*. Prentice-Hall, Hoboken.

Paaby A. B., and M. V. Rockman, (2013) 'The many faces of pleiotropy'. *Trends Genet.* 29: 66–73. https://doi.org/10.1016/j.tig.2012.10.010

Pagel M., and R. A. Johnstone, (1992). 'Variation across species in the size of the nuclear genome supports the junk-DNA explanation for the C-value paradox'. *Proc. R. Soc. B Biol. Sci.* 249: 119–124. https://doi.org/10.1098/rspb.1992.0093

Paley W., (1785 [2002]). *The Principles of Moral and Political Philosophy*. R. Faulder, London [Liberty Fund, Indianapolis].

Paley W., (1794 [1877]). *A View of the Evidences of Christianity*. R. Faulder, London [Ward, Lock and Co., London, New York and Melbourne].

Paley W., (1802 [1809]). *Natural Theology, or Evidences of the Existence and Attributes of the Deity*. Wilks & Taylor, London [J. Faulder, London]. http://darwin-online.org.uk/content/frameset?itemID=A142&pageseq=1&viewtype=text

Park T., P. H. Leslie, and D. B. Mertz, (1964). 'Genetic strains and competition in populations of Tribolium'. *Physiol. Zool.* 37: 97–162. https://doi.org/10.1086/physzool.37.2.30152328

Parker J. E., S. P. Knowler, C. Rusbridge, et al., (2011). 'Prevalence of asymptomatic syringomyelia in Cavalier King Charles spaniels'. *Vet. Rec.* 168: 667. https://doi.org/10.1136/vr.d1726

Parvinen K., (2005). 'Evolutionary suicide'. *Acta Biotheor.* 53: 241–264. https://doi.org/10.1007/s10441-005-2531-5

Patel M., S. A. West, and J. M. Biernaskie, (2020). 'Kin discrimination, negative relatedness, and how to distinguish between selfishness and spite'. *Evol. Lett.* 1–8. https://doi.org/10.1002/evl3.150

Pellestor F., B. Andreo, T. Anahory, et al., (2006). 'The occurrence of aneuploidy in human: lessons from the cytogenetic studies of human oocytes'. *Eur. J. Med. Genet.* 49: 103–116. https://doi.org/10.1016/j.ejmg.2005.08.001

Pichot, C., Borrut, A., and El Maâtaoui, M., (1998). 'Unexpected DNA content in the endosperm of *Cupressus dupreziana* A. Camus seeds and its implications in the reproductive process'. *Sex. Plant Reprod.* 11: 148–152.

Pichot, C., El Maâtaoui, M., S. Raddi, et al., (2001). 'Surrogate mother for endangered Cupressus'. *Nature.* 412: 39.

Pierotti R., (1980). 'Spite and Altruism in Gulls'. *Am. Nat.* 115: 290–300. https://doi.org/10.1086/283561

Pigou A. C., (1920 [2013]). *The Economics of Welfare*. Macmillan and Co., London [Palgrave Macmillan, Basingstoke].

Pointer M. D., M. J. G. Gage, and L. G. Spurgin, (2021). 'Tribolium beetles as a model system in evolution and ecology'. *Heredity* (Edinb). 126: 869–883. https://doi.org/10.1038/s41437-021-00420-1

Popper K., (1935 [2002]). *The Logic of Scientific Discovery*. Julius Springer, Hutchinson & Co., Vienna [Routledge Classics, London and New York].

Popper K. R., (1963 [2002]). *Conjectures and Refutations: The Growth of Scientific Knowledge*. Routledge & Kegan Paul, Oxford [Basics Books, New York and London].

Popper K. R., (1972 [1974]). *Objective Knowledge: An Evolutionary Approach*. Oxford University Press, London.

Pracana R., I. Levantis, C. Martínez-Ruiz, et al., (2017). 'Fire ant social chromosomes: differences in number, sequence and expression of odorant binding proteins'. *Evol. Lett.* 1: 199–210. https://doi.org/10.1002/evl3.22

Price T. A. R., G. D. D. Hurst, and N. Wedell, (2010). 'Polyandry prevents extinction'. *Curr. Biol.* 20: 471–475. https://doi.org/10.1016/j.cub.2010.01.050

Price T. A. R., R. Verspoor, and N. Wedell, (2019). 'Ancient gene drives: An evolutionary paradox'. *Proc. R. Soc. B Biol. Sci.* 286. https://doi.org/10.1098/rspb.2019.2267

Price T. A. R., N. Windbichler, R. L. Unckless, et al., (2020). 'Resistance to natural and synthetic gene drive systems'. *J. Evol. Biol.* 33: 1345–1360. https://doi.org/10.1111/jeb.13693

Pross A., (2012 [2016]). *What is Life?: How Chemistry becomes Biology*. Oxford University Press, New York.

Pye R. J., G. M. Woods, and A. Kreiss, (2016). 'Devil facial tumor disease'. *Vet. Pathol.* 53: 726–736. https://doi.org/10.1177/0300985815616444

Queller D. C., (1994). 'Genetic relatedness in viscous populations'. *Evol. Ecol.* 8: 70–73. https://doi.org/10.1007/BF01237667

Queller D. C., (1997). 'Cooperators Since Life Began'. *Q. Rev. Biol.* 72: 184–188. https://doi.org/10.1086/419766

Queller D. C., E. Ponte, S. Bozzaro, et al., (2003) 'Single-Gene Greenbeard Effects in the Social Amoeba *Dictyostelium discoideum*'. *Science* 299: 105–106. https://doi.org/10.1126/science.1077742

Rankin D. J., K. Bargum, and H. Kokko, (2007). 'The tragedy of the commons in evolutionary biology'. *Trends Ecol. Evol.* 22: 643–651. https://doi.org/10.1016/j.tree.2007.07.009

Ratnieks F. L. W., and H. K. Reeve, (1992). 'Conflict in single-queen hymenopteran societies: the structure of conflict and processes that reduce conflict in advanced eusocial species'. *J. Theor. Biol.* 158: 33–65. https://doi.org/10.1016/S0022-5193(05)80647-2

Rebbeck C. A., R. Thomas, M. Breen, et al., (2009). 'Origins and evolution of a transmissible cancer'. *Evolution* (NY) 63: 2340–2349. https://doi.org/10.1111/j.1558-5646.2009.00724.x

Rice W. R., (2013). 'Nothing in genetics makes sense except in light of genomic conflict'. *Annu. Rev. Ecol. Evol. Syst.* 44: 217–237. https://doi.org/10.1146/annurev-ecolsys-110411-160242

Ridley M., and A. Grafen, (1981). 'Are green beard genes outlaws?' *Anim. Behav.* 29: 954–955. https://doi.org/10.1016/S0003-3472(81)80034-6

Ridley M., (1996 [1997]). *The Origins of Virtue*. Viking, London [Penguin Books, London].

Ridley M., (2000). *Mendel's Demon: Gene Justice and the Complexity of Life*. Weidenfeld & Nicolson, London.

Ridley M., (1993 [2004]). *Evolution*. Blackwell Science, Oxford.

Roff D. A., (2001 [2002]). *Life History Evolution*. Sinauer Associates, Sunderland.

Roger A. J., S. A. Muñoz-Gómez, and R. Kamikawa, (2017). 'The origin and diversification of mitochondria'. *Curr. Biol.* 27: R1177–R1192. https://doi.org/10.1016/j.cub.2017.09.015

Roth G., K. C. Nishikawa, and D. B. Wake, (1997). 'Genome size, secondary simplification, and the evolution of the brain in salamanders'. *Brain. Behav. Evol.* 50: 50–59. https://doi.org/10.1159/000113321

Rothstein S. I., and D. P. Barash, (1983). 'Gene conflicts and the concepts of outlaw and sheriff alleles'. *J. Soc. Biol. Syst.* 6: 367–379. https://doi.org/10.1016/S0140-1750(83)90156-2

Ryle G., (1949 [1970]). *The Concept of Mind*. Hutchinson & Co., London [Peregrine Books, London].

Sandler S., and E. Novitski, (1957). 'Meiotic drive as an evolutionary force'. *Am. Nat.* 857: 105–110. https://doi.org/10.1086/281969

Schimenti J. C., J. L. Reynolds, and A. Planchart, (2005). 'Mutations in Serac1 or Synj2 cause proximal t haplotype-mediated male mouse sterility but not transmission ratio distortion'. *Proc. Natl. Acad. Sci. USA.* 102: 3342–3347. https://doi.org/10.1073/pnas.0407970102

Sclavi B., and J. Herrick, (2019). 'Genome size variation and species diversity in salamanders'. *J. Evol. Biol.* 32: 278–286. https://doi.org/10.1111/jeb.13412

Scofield V. L., J. M. Schlumpberger, L. A. West, et al., (1982). 'Protochordate allorecognition is controlled by a MHC-like gene system'. *Nature* 295: 499–502. https://doi.org/10.1038/295499a0

Scott T. W., and S. A. West, (2019). 'Adaptation is maintained by the parliament of genes'. *Nat. Commun.* 10: 1–13. https://doi.org/10.1038/s41467-019-13169-3

Shorter J. R., and O. Rueppell, (2012). 'A review on self-destructive defense behaviors in social insects'. *Insectes Soc.* 59: 1–10. https://doi.org/10.1007/s00040-011-0210-x

Silver L. M., (1993). 'The peculiar journey of a selfish chromosome: mouse t haplotypes and meiotic drive'. *Trends Genet.* 9: 250–254. https://doi.org/10.1016/0168-9525(93)90090-5

Simpson G. G., (1966). 'The biological nature of man'. *Science* 152: 472–478. https://doi.org/10.2307/2010269

Sober E., (1984 [1993]). *The Nature of Selection: Evolutionary Theory in Philosophical Focus.* MIT Press, Cambridge [University of Chicago Press, Chicago].

Southwood R., (2003 [2004]). *The Story of Life.* Oxford University Press, New York.

Spiegelman S., (1967). 'An in vitro analysis of a replicating molecule'. *Am. Sci.* 55: 221–264.

Spiegelman S., (1971). 'An approach to the experimental analysis of precellular evolution'. *Q. Rev. Biophys.* 4: 213–253. https://doi.org/10.1017/s0033583500000639

Stevens L., (1989). 'The genetics and evolution of cannibalism in flour beetles (genus Tribolium)'. *Evolution* (NY) 43: 169–179. https://doi.org/10.1111/j.1558-5646.1989.tb04215.x

Strassmann J. E., Y. Zhu, and D. C. Queller, (2000). 'Altruism and social cheating in the social amoeba *Dictyostelium discoideum*'. *Nature* 408: 965–967. https://doi.org/10.1038/35050087

Strassmann J. E., and D. C. Queller, (2010). 'The social organism: congresses, parties, and committees'. *Evolution* (NY) 64: 605–616. https://doi.org/10.1111/j.1558-5646.2009.00929.x

Sulak M., L. Fong, K. Mika, et al., (2016). 'TP53 copy number expansion is associated with the evolution of increased body size and an enhanced DNA damage response in elephants'. *Elife* 5: 1–30. https://doi.org/10.7554/eLife.11994

Sutter A., and A. K. Lindholm, (2015). 'Detrimental effects of an autosomal selfish genetic element on sperm competitiveness in house mice'. *Proc. R. Soc. B Biol. Sci.* 282: 20150974. https://doi.org/10.1098/rspb.2015.0974

Taylor P. D., (1992). 'Altruism in viscous populations—an inclusive fitness model'. *Evol. Ecol.* 6: 352–356. https://doi.org/10.1007/BF02270971

Tollis M., A. M. Boddy, and C. C. Maley, (2017). 'Peto's Paradox: how has evolution solved the problem of cancer prevention?' *BMC Biol.* 15: 1–5. https://doi.org/10.1186/s12915-017-0401-7

Trible W., and K. G. Ross, (2015). 'Chemical communication of queen supergene status in an ant'. *J. Evol. Biol.* 29: 502–513. https://doi.org/10.1111/jeb.12799

Trivers R. L., (1985). *Social Evolution*. Benjamin/Cummings Publishing Co., Menlo Park, CA.

Unckless R. L., A. G. Clark, and P. W. Messer, (2017). 'Evolution of resistance against CRISPR/Cas9 gene drive'. *Genetics* 205: 827–841. https://doi.org/10.1534/genetics.116.197285

UNDESA, (2022). World Population Prospects 2022. https://www.un.org/development/desa/pd/.

Vincze O., F. Colchero, J. F. Lemaître, et al., (2022). 'Cancer risk across mammals'. *Nature* 601: 263–267. https://doi.org/10.1038/s41586-021-04224-5

Vinogradov A. E., (2003). 'Selfish DNA is maladaptive: evidence from the plant Red List'. *Trends Genet.* 19: 609–614. https://doi.org/10.1016/j.tig.2003.09.010

Vinogradov A. E., (2004). 'Genome size and extinction risk in vertebrates'. *Proc. R. Soc. B Biol. Sci.* 271: 1701–1705. https://doi.org/10.1098/rspb.2004.2776

Voltaire, (1759 [1969]). *Candide and Other Tales*. Cramer, Marc-Michel Rey, Jean Nourse, Lambert, and others [Heron Books, Bristol].

Vries H. de, (1901–1903 [(1910)]). *The Mutation Theory: Experiments and Observations on the Origin of Species in the Vegetable Kingdom*. Leipzig, Veit & Company, Leipzig [Open Court Publishing, London]. https://www.biodiversitylibrary.org/item/16402#page/17/mode/1up

Wade M. J., (1978). 'A critical review of the models of group selection'. *Q. Rev. Biol.* 53: 101–114. https://doi.org/10.1086/410450

Wade M. J., (1980). 'An experimental study of kin selection'. *Evolution* (NY) 34: 844–855. https://doi.org/10.2307/2407991

Wagner G. P., and J. Zhang, (2011) 'The pleiotropic structure of the genotype-phenotype map: the evolvability of complex organisms'. *Nat. Rev. Genet.* 12: 204–213. https://doi.org/10.1038/nrg2949

Wakap S. N., D. M. Lambert, A. Olry, et al., (2020). 'Estimating cumulative point prevalence of rare diseases: analysis of the Orphanet database'. *Eur. J. Hum. Genet.* 28: 165–173. https://doi.org/10.1038/s41431-019-0508-0

Wald A., (1943). 'A method of estimating plane vulnerability based on damage of survivors'. *Cent. Nav. Anal.* CRC 432: 1–89.

Wallace A. R., (1889). *Darwinism: An Exposition of the Theory of Natural Selection with some of Its Applications*. Macmillan & Co., London and New York. http://darwin-online.org.uk/converted/Ancillary/1889_Darwinism_S724/1889_Darwinism_S724.html

Wallace B., (1975). 'Hard and soft selection revisited'. *Evolution* (NY) 29: 465–473. https://doi.org/10.1111/j.1558-5646.1975.tb00836.x

Waltz E. C., (1981). 'Reciprocal altruism and spite in gulls: a comment'. *Am. Nat.* 118: 588–592. https://doi.org/10.1126/science.26.678.918

Wang J., Y. Wurm, M. Nipitwattanaphon, et al., (2013) 'A Y-like social chromosome causes alternative colony organization in fire ants'. *Nature.* 493: 664–668. https://doi.org/10.1038/nature11832

Wang, L., F. S. Wang, and M. E. Gershwin, (2015). 'Human autoimmune diseases: a comprehensive update'. *J. Intern. Med.* 278: 369–395. https://doi.org/10.1111/joim.12395

Watson J. D., and F. H. C. Crick, (1953). 'Molecular structure of nucleic acids: a structure for deoxyribose nucleic acid'. *Nature.* 171: 737–738. https://doi.org/10.1038/171737a0

Webb C., (2003). 'A complete classification of Darwinian extinction in ecological interactions'. *Am. Nat.* 161: 181–205. https://doi.org/10.1086/345858

Wenseleers T., A. G. Hart, F. L. W. Ratnieks, et al., (2004). 'Queen execution and caste conflict in the stingless bee *Melipona beecheii*'. *Ethology.* 110: 725–736. https://doi.org/10.1111/j.1439-0310.2004.01008.x

Wenseleers T., and F. L. W. Ratnieks, (2004). 'Tragedy of the commons in Melipona bees'. *Proc. R. Soc. B, Biol. Sci.* 271: S310–312. https://doi.org/10.1098/rsbl.2003.0159

Werren J. H., U. Nur, and C. I. Wu, (1988). 'Selfish genetic elements'. *Trends Ecol. Evol.* 3: 297–302. https://doi.org/10.1016/0169-5347(88)90105-X

West S. A., I. Pen, and A. S. Griffin, (2002). 'Cooperation and competition between relatives'. *Science* 296: 72–75. https://doi.org/10.1126/science.1065507

West S. A., and A. Buckling, (2003). 'Cooperation, virulence and siderophore production in bacterial parasites'. *Proc. Biol. Sci.* 270: 37–44. https://doi.org/10.1098/rspb.2002.2209

West S. A., A. S. Griffin, and A. Gardner, (2007a). 'Social semantics: altruism, cooperation, mutualism, strong reciprocity and group selection'. *J. Evol. Biol.* 20: 415–432. https://doi.org/10.1111/j.1420-9101.2006.01258.x

West S. A., A. S. Griffin, and A. Gardner, (2007b). 'Evolutionary explanations for cooperation'. *Curr. Biol.* 17: R661–R672. https://doi.org/10.1016/j.cub.2007.06.004

West S. A., (2009). *Sex Allocation*. Princeton University Press, Princeton, NJ.

West S. A., and A. Gardner, (2010). 'Altruism, spite, and greenbeards'. *Science* 327: 1341–1344. https://doi.org/10.1126/science.1178332

West S. A., and A. Gardner, (2013). 'Adaptation and inclusive fitness'. *Curr. Biol.* 23: R577–584. https://doi.org/10.1016/j.cub.2013.05.031

Williams G. C., and D. C. Williams, (1957). 'Natural selection of individually harmful social adaptations among sibs with special reference to social insects'. *Evolution* (NY) 11: 32. https://doi.org/10.2307/2405809

Williams G. C., (1966 [1996]). *Adaptation and Natural Selection: A Critique of some Evolutionary Thought*. Princeton University Press, Princeton, NJ.

Williams G. C., (1975 [1977]). *Sex and Evolution*. Princeton University Press, Princeton, NJ.

Williams G. C., (1992) *Natural Selection: Domains, Levels, and Challenges*. Oxford University Press.

Wilson D. S., (1975a). 'A theory of group selection'. *Proc. Natl. Acad. Sci.* 72: 143–146. https://doi.org/10.1073/pnas.72.1.143

Wilson E. O., (1975b [2000]). *Sociobiology: The New Synthesis*. Harvard University Press, Cambridge, MA.

Wilson E. O., and B. Hölldobler, (1990 [1998]). *The Ants*. Harvard University Press, Cambridge, MA.

Wilson D. S., G. B. Pollock, and L. A. Dugatkin, (1992). 'Can altruism evolve in purely viscous populations?' *Evol. Ecol.* 6: 331–341. https://doi.org/10.1007/BF02270969

Wood R. J., and M. E. Newton, (1991). 'Sex-ratio distortion caused by meiotic drive in mosquitoes'. *Am. Nat.* 137: 379–391. https://doi.org/10.1086/285171

Wright S., (1922). 'Coefficients of relationship and inbreeding'. *Am. Nat.* 56: 330–338. https://doi.org/10.1086/279872

Wright S., (1931). 'Evolution in Mendelian populations'. *Genetics.* 16: 97–159. https://doi.org/10.1007/BF02459575

Wright S., (1945). 'Tempo and mode in evolution: a critical review'. *Ecology.* 26: 415–419. https://doi.org/10.2307/1931666

Wright S., (1951). 'The genetical structure of populations'. *Ann. Eugen.* 15: 323–354. https://doi.org/10.2307/2407273

Wright S., (1969). *Evolution and the Genetics of Populations. Volume 2: The Theory of Gene Frequencies*. University of Chicago Press, Chicago.

Wynne-Edwards V. C., (1962 [1972]). *Animal Dispersion in Relation to Social Behaviour*. Oliver & Boyd, Edinburgh [Hafner Publishing Co., New York and London].

Wynne-Edwards V. C., (1986). *Evolution through Group Selection*. Blackwell Science, Oxford.

Yahalomi D., S. D. Atkinson, M. Neuhof, et al., (2020). 'A cnidarian parasite of salmon (Myxozoa: Henneguya) lacks a mitochondrial genome'. *Proc. Natl. Acad. Sci. U. S. A.* 117: 5358–5363. https://doi.org/10.1073/pnas.1909907117

Yanai I., and M. Lercher, (2016). *The Society of Genes*. Harvard University Press, Cambridge, MA.

Zahavi A., (1975). 'Mate selection—a selection for a handicap'. *J. Theor. Biol.* 53: 205–214. https://doi.org/10.1016/0022-5193(75)90111-3

Zahavi, A., and A. Zahavi, (1997 [1999]). *The Handicap Principle: A Missing Piece of Darwin's Puzzle.* Oxford University Press, New York.

Zanders S. E., and R. L. Unckless, (2019). 'Fertility costs of meiotic drivers'. *Curr. Biol.* 29: R512–R520. https://doi.org/10.1016/j.cub.2019.03.046

Zhang H., and S. Chen, (2016) 'Tag-mediated cooperation with non-deterministic genotype–phenotype mapping'. *Europhys. Lett.* 113: 28008. https://doi.org/10.1209/0295-5075/113/28008

Zimmering S., L. Sandler, and B. Nicoletti, (1970). 'Mechanisms of meiotic drive'. *Annu. Rev. Genet.* 4: 409–436. https://doi.org/10.1146/annurev.ge.04.120170.002205.

Index

For the benefit of digital users, indexed terms that span two pages (e.g., 52–53) may, on occasion, appear on only one of those pages.